Thomas C. Martin, Joseph Sachs

Electrical Boats and Navigation

Thomas C. Martin, Joseph Sachs

Electrical Boats and Navigation

ISBN/EAN: 9783337412210

Printed in Europe, USA, Canada, Australia, Japan

Cover: Foto ©berggeist007 / pixelio.de

More available books at **www.hansebooks.com**

ELECTRICAL BOATS

AND

NAVIGATION

BY

THOMAS COMMERFORD MARTIN

Past President American Institute Electrical Engineers. Editor The Electrical Engineer.

AND

JOSEPH SACHS

Member N. Y. Electrical Society and Associate Member A. I. E. E.

NEW YORK
C. C. SHELLEY, PUBLISHER,
10 & 12 COLLEGE PLACE.

1894.

PREFACE.

IT is believed that the contents of this volume will be of interest to a wide circle of readers, many of whom have not hitherto paid attention to the rapid advances made in the application of electric power to the purposes of navigation. The industry of equipping and operating electrical boats is still new ; but, as will be seen, a great deal of instructive experimental work has been done, and in many instances marked practical success has been attained. No attempt has been made here to give the details of every electric craft built up to date, but a certain number are dealt with as illustrative of the history of the art and of its evolution. The book also embodies a large quantity of data on the subject not brought together before. It has sometimes been found that an art passing through its earlier stages of perfection is helped by the appearance of literature on its new problems and conditions ; and should any stimulus to electrical navigation be given by this book, the authors will feel rewarded for their trouble.

Mr. Martin has taken charge of the preparation of the book as a whole. Mr. Sachs has particularly applied himself to the uses of electricity in canal boat propulsion and similar work, including the data on screws, resistance, etc. The canal section is, in fact, primarily based upon a lecture delivered by Mr. Sachs before the New York Electrical Society.

Fuller details could have been given with regard to some of the classes of boats here discussed, whether automobile or deriving live current by wire from a central source of supply; but it is hoped that the book, as it stands, may subserve the needs of the present moment, while the art is still in a transition state. References are given in various places to authorities that may be consulted with benefit, as to detail apparatus.

The authors wish to acknowledge their indebtedness to Mr. J. C. Chamberlain and Mr. F. Reckenzaun for suggestions and assistance; and to Mr. Joseph Wetzler for collaboration in passing the book through the press.

THOMAS COMMERFORD MARTIN.
JOSEPH SACHS.

NEW YORK CITY. September, 1894.

CONTENTS.

ELECTRICAL BOATS AND NAVIGATION.

CHAPTER I.

ELECTRICAL BOATS : HISTORICAL AND INTRODUCTORY.—
PRIMARY BATTERY BOATS.

1. The success of electric locomotion on land is now well established, and thousands of electric cars are to-day in use on the streets, on elevated roads, in tunnels beneath cities, and across country. But while electric traction has thus been advancing rapidly, electrical navigation has also made progress more quietly, as an art and an industry ; and it is believed that the time has arrived when the subject of electrical boats, for a large variety of services, may be discussed with pleasure and profit. As the succeeding pages will show, electrical navigation divides under two main heads, namely, that which includes automobile or independent boats carrying their own power, and that which includes dirigible or dependent boats deriving current all the time from a fixed point on shore. In some instances the characteristics of the two groups interlap, but the arrangement of the chapters in this volume has been made in accordance with the classification suggested by such a natural difference in function and performance.

Nearly sixty years ago—1838—Prof. Jacobi, a distinguished physicist and electrician residing at St. Petersburg, made a demonstration on the Neva of what could be accomplished with the electric motor as a means of propulsion for boats. Towards the expenses of the experiment, the Emperor Nicholas contributed $12,000. Jacobi employed a form of motor which had already attracted considerable notice, having been brought before the Académie des Sciences of Paris in 1834. This attempt is the first of the kind on record, and is another proof of the originality and versatility in electrical investigation that made Jacobi one of the discoverers of the great modern art of electroplating. Just what led the Emperor to take an interest in the application of electricity to marine propulsion and spend this money on it is not known, but it is supposed that these tests, like others, then, in Europe, were due to

Fig. 1.—Jacobi's Electric Boat Motor of 1838-9.

the exhibitions made in London by Thomas Davenport, the Vermont blacksmith, of his electric motors. The Jacobi boat on the Neva derived its current from primary batteries, and the motor propelled it at a speed which never exceeded 3 miles an hour. Both Grove and Daniell cells were used in the first and succeeding experiments. The motor (Fig. 1) was geared to paddle wheels. The boat was 28 feet long, 7 feet wide and 2 feet 8 inches in draught. She carried as many as 14 passengers.

Since that time and the later work in England of Robert Hunt; of G. E. Dering, in 1856; and the Count de Moulins in France in 1866, there have been a great many other attempts

FIG. 2.—TROUVÉ'S ELECTRIC BOAT ON THE SEINE.

FIG. 2a.—TROUVÉ'S BOAT FOR FIVE PASSENGERS.

to apply primary batteries to the propulsion of boats, but
the experiments have nowhere resulted in lasting success.
Perhaps the most notable instance is the work done by the
ingenious French inventor, Trouvé, who at the Paris Exposition of 1881 showed a small boat (Fig. 2) of somewhat
novel construction.   He employed a motor, placed imme-

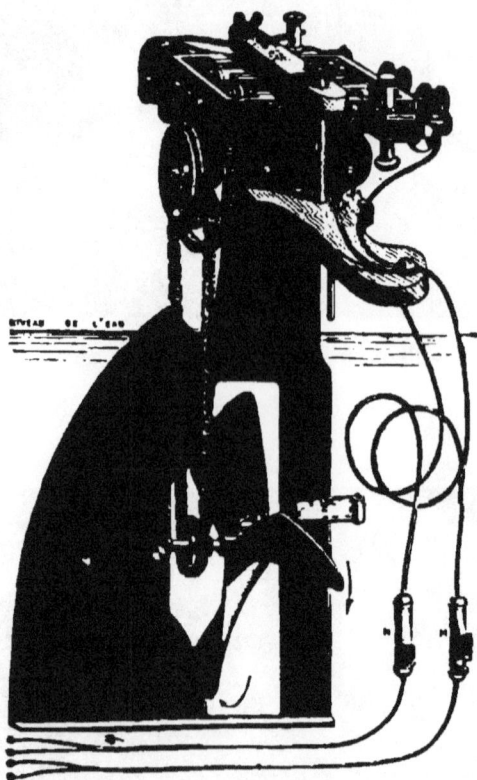

FIG. 2b.—TROUVÉ'S MOTOR AND SCREW.

diately over the rudder and driving by means of sprocket
cogwheel and chains a three-bladed screw, carried in the
rudder frame (Fig. 2b). Current was conveyed to this motor
by means of two little flexible cables from the batteries;
and the cables served also as yoke lines. The batteries
were of the bichromate of potash type, with the plates so

arranged as to be raised from or lowered into the solution. Trouvé also tried Planté storage cells, but gave them up. With a load of four or five passengers, this boat would attain a speed of about 3½ miles an hour. Fig. 2a shows a similar Trouvé boat of the same period.

Other workers here and in Europe have made various boats for primary battery use, and some are probably running at this moment. There is in reality no reason why

FIG. 3.—THE PRIMARY BATTERY LAUNCH "ELECTRIC" ON THE POTOMAC.

they should not be numerous in a day when every newborn genius dabbles in electricity and when any form of electric locomotion has undoubted fascination for the public. An ordinary primary battery is not difficult to manage, and if a good form of solution is used, such as the bichromate of potash, a long run, say of four or five consecutive hours, may be made by a light boat, at a cost for motive power not exceeding 10 cents an hour, and even less.

2. In order to show what may be done with primary battery boats, we will illustrate and describe in passing,

one with which many trips have been made on the Potomac at Washington, and one whose size marks her as the most pretentious of her class.

This launch (Fig. 3) which was built by the Naphtha Launch Co., of Morris Dock, N. Y., and A. Glose & Son, New York, is 21 feet in length and 5 feet beam. She draws 18 inches of water with battery and motor, which are thus, it will be observed, barely sufficient to serve as ballast, instead of being an encumbering dead weight. The propeller is 17 inches in diameter with three blades, and makes 250 revolutions per minute. To the propeller is geared a small Riker motor wound for 50 volts and 30 amperes, and weighing 240 pounds. The gear wheels are respectively 2¾ inches and 6 inches. The launch has averaged 7 miles an hour.

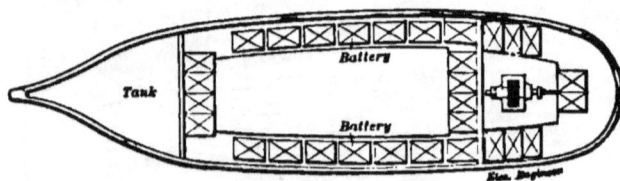

FIG. 4.—PLAN OF PRIMARY BATTERY LAUNCH "ELECTRIC."

The battery equipment is very interesting. It consists of 60 Hanson cells grouped as shown in the engraving, Fig. 4, with 30 cells on each side, and so connected that half the required current comes from each row of cells. Each battery box holds two cells. Each cell is 11 x 12 x 3 inches and in each there is a porous cup 9 x 11 x 1 inch, in which is suspended a carbon plate 7 x 11 x ¼ inch. Outside the porous cup are suspended two zinc plates each 7 x 10 x ¼ inch, the surface being thoroughly exposed to the action of the solution. In the porous cup are three pints of strong solution and in the outer cell are six pints of weak solution. The total cost of the solution, including depolarizer, for the 60 cells, is $1.60. From one charge of this battery, the yield has, it is said, been 180 ampere hours. The cell gives an average of 1¾ volts, and will yield 50 to 60 amperes on short circuit. The zincs are amalgamated before using, and again after 24 hours' use. The average consumption of zinc is less than one pound per hour when the battery is in use; and when it is not in use the local action is not sufficient to render it necessary to remove the zincs. Each double cell weighs 40 pounds, or 1,200 pounds for the 60 cells, when the zincs are new. The battery cells are of wood treated with a special preparation making them acid-proof.

' From one charge of the solution, the boat gets a good 12 hours' run. In the tank shown at the bow of the boat, is carried extra solution enough for three more charges.

FIGS. 5 AND 5a — VAUHAN-SHERRIN PRIMARY BATTERY BOAT.

One man can recharge the whole 60 cells in 1½ hours. Of course the 12 hours of service can be extended over several days, a few hours at a time. The current is thrown on or off by a simple switch that anyone can control. The "Electric" seats 12 persons easily.

3. One of the most noteworthy of the primary battery boats built in England is that of Mr. Vauhan-Sherrin, (Figs. 5 and 5a) launched a few years ago and tried during 1890-91, although no recent reports have been made public. The primary battery used by Mr. Sherrin is a two-fluid form, in which the anodes are of sheet zinc, and the

cathodes are of carbon, specially prepared. In each cell there are three fixed cathodes and two replaceable anodes. Very light plates are used, and the particular construction adopted permits these to be placed very close together, so that the internal resistance is small. The outer cells are of gutta-percha, and in them are embedded the porous cells which surround the anodes. The liquid used in the inner cells is simply water; that placed in the outer cells around the carbon cathodes is a depolarizing liquid of special composition, capable of being produced at a low cost. In one of the tests made by Professor Silvanus P. Thompson, one of these cells gave out an average current of 8.75 amperes for five consecutive hours, with an average electromotive force of 1.88 volts, although the cell was only about half filled at starting. Professor Thompson says that he knows of no battery, primary or secondary, which, for a given gross weight of cell, will yield as great an output, while the economy of zinc is remarkable, the consumption being close to the theoretical limit. The net cost of electric energy from such cells is estimated at 18 to 20 cents per Board of Trade unit of 1,000 watt hours.

The motor is a modified two pole Gramme machine, having the field magnets constructed in a special manner, which, while maintaining great mechanical strength, admits of perfect lamination of the iron. It is well designed and constructed, and when properly set, is free from sparking at the commutator. It is also light, a one horse-power motor representing only 62 lbs. of dead weight.

The inventor can carry his battery composition in the form of a paste, which, by merely mixing with water, forms at once a fresh charge for his batteries. One writer describing this boat reports having seen a launch 40 feet long, belonging to Mr. Sherrin, and fitted under his system with a capacity for a 600 mile continuous run.

4. The statement has been made of another primary battery boat built in 1892, to run on the Schuylkill River, that she has made a trip from New York to Philadelphia, or about 90 miles, on one charge of solution; but the report lacks "corroborative detail." In the meantime, in view of the many attempts still likely to be made in this direc-

tion, it is worth while to point out that any figures given
as to the marvelous capacity of primary batteries to do a
great deal of work for nothing and to yield a profit on the
residual by-products should be received with a fair amount
of caution. A valuable discussion of this subject will be
found in a paper by Prof. Francis B. Crocker, of Columbia
College, read before the American Institute of Electrical
Engineers, in 1888, on "The Possibilities and Limitations
of Chemical Generators of Electricity." This paper has
never been used as part of the prospectus for floating the
stock of a new primary battery company. We may quote
the conclusion of Prof. Crocker's remarks:—"I am by no
means a skeptic in regard to chemical generators of electricity; the possibilities are very, very great as I have
shown. But these possibilities do not seem to have been
brought to reality in a very perfect manner as yet. But
batteries even in the imperfect state in which they exist
to-day have their useful and legitimate function. A
Leclanché cell is exceedingly well adapted to ringing electric bells intermittently and to telephone work; and
gravity batteries have long done good service for tele-
graphic purposes.

"But when it comes to developing any considerable
amount of actual power, then the limitations become apparent. When we remember that battery electricity will certainly cost in practice 50 cents per horse-power hour, since
the materials alone cost 25 cents, and that dynamo electricity only costs 2 cents per horse-power hour, the claims of
some primary battery electric lighting promoters show up
in their true light. As a luxury, of course, it makes very
little difference what it costs, but even then people soon
tire of paying very high prices for that kind of a luxury.
For small electric lighting and small power in special cases,
batteries are useful, particularly where no other source of
current is available. A physician or dentist to whom a
horse-power hour may be worth hundreds of dollars could
almost afford to use a chloride of silver battery and throw
away the silver."

5. These cautionary remarks may fitly close a chapter
on primary battery boats. If a good battery be obtained

for which no extravagant claims are made, and if the boat
be equipped with almost any one of the large number of
small motors on the market, its owner should be able to
derive many pleasant hours from its possession, with absolute, or almost entire, respite from the fatigue of rowing or
the care of tending sails.   To invalids and lazy people this
is quite a consideration.   Any ordinary rowboat can be
converted into an electric boat.   The apparatus complete
for a 12 foot boat, to make a speed not greatly exceeding
4 miles an hour will run from $150 to $250, placing the

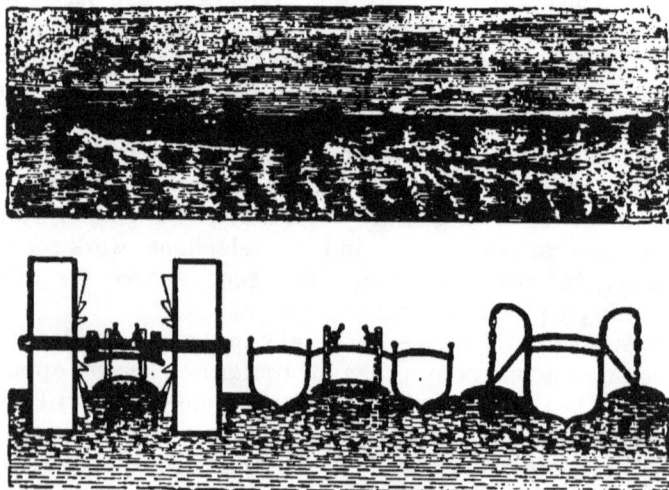

FIGS. 6 AND 6a.—TROUVÉ'S RAFT BATTERY FOR SEA-GOING SHIPS.

equipment within the power of persons of moderate means.
To this sum must be added the cost of renewal of zincs and
solution, for which 50 cents a day would be a fair allowance.

It will be gathered from what has been said that the primary battery boat is looked upon as one for individual
and casual use rather than for such units and heavy service
as would constitute in magnitude and importance a new
development of electrical engineering.   At the same time,
we must not forget to mention that, returning to ideas ad-
vanced at least fifty years ago, M. Trouvé has suggested that
large electric boats for sea-going purposes may be propelled

by primary batteries, the elements of which are immersed in
the salt water of the ocean as an electrolyte. For the sake
of the curious, we here reproduce sketches (Figs. 6 and 6a)
from the history of M. Trouvé's inventions,' showing an
electric ship to which is attached a "raft-battery," the
cables from which carry the current to the motor on the
ship, the raft consisting of copper-zinc couples, so grouped
as to give the desired voltage and amperage. The detail
drawings illustrate other plans of supporting the elements,
with arrangements for greater or less submersion in the
sea. M. Trouvé assumes that from such a battery he might
get 60 watts per square metre of surface of zinc, both sides
of the plates being utilized ; and that a raft 100 metres
long and 16 metres wide would by a proper disposition of
the plates, dipping 4 metres into the sea, yield him 5,120
horse-power (French) on the same basis of calculation'. On
arriving in port, it would be much easier, he thinks, to
renew this "beautiful magazine of energy," by simply taking another "raft" off the wharf or out of dry dock, than
to recoal a steamship of corresponding capacity. It is
natural to express the wish that some experiments might
be made in this direction, at least on a small scale. That
we shall ever see a large ship laboriously hauling an acre
or two of such batteries through the surf, is almost too
much to expect. Many persons would prefer to try generation of current by thermopiles or pyromagnetic means.

CHAPTER II.

STORAGE BATTERY BOATS.—PRELIMINARY : SINGLE LAUNCHES.

6. The electrical propulsion of boats was one of the many industries which had to await the perfection of means of generating current cheaply. Its success depended also upon improvements in motors and in no small degree upon the introduction of the modern storage battery. Not only did the fumes from Jacobi's batteries drive away the spectators on the banks, but his apparatus was as a whole the most costly and inefficient that could well be conceived as likely to drive away capital, although at the time of its use it was practically the best to be had. It is not less true that each of the later experiments down to 1881 represented the hopes raised by some new departure in the electric arts, hopes that could not be realized, however, under the conditions that had to be encountered.

The great advance made in the production of the dynamo-electric machine had an immediate influence upon the electric motor, while at the same time the fact that large generating machines could be run by engines and turbines stimulated experiment with storage batteries in which some of their current might be accumulated. One of the most noteworthy results of this interaction and parallel evolution is the storage battery boat of to-day, which is already so successful and which promises to add so greatly to the comfort, convenience and welfare of all whose pleasure or business takes them upon the water.

In his boat at Paris, Trouvé bridged the gap between the old practice and the new, but the first serious attempt in the direction of utilizing stored current was made in 1882, when the late Anthony Reckenzaun, the brilliant young Austrian electrical engineer, who died all too soon, designed a launch called "Electricity," for the Electric Power

Storage Company, of London. She was a boat of good size, being 25 feet in length and about 5 feet in beam. Her draught of water was 1 foot 9 inches forward, and 2 feet 6 inches aft. During the earlier experiments with her, current was furnished from 45 accumulators of the Sellon-Volckmar (Faure) type. These were connected to two Siemens motors which operated singly or jointly, and which communicated power by countershafting to a 22-inch propeller screw. The boat would carry 12 passengers, and

FIG. 7.—THE "VOLTA" ELECTRIC LAUNCH.

made as high a speed, it is said, as 8 knots an hour against the Thames tide.

7. The next important step was taken in 1883, when the Yarrows, of England, exhibited at the Vienna Electrical Exhibition a launch 46 feet in length, capable of carrying 50 passengers, and able to maintain a speed of 8 or 9 miles an hour. Current was furnished by 70 accumulators stowed away under the floor of the launch. The motor, of the Siemens type, drove directly, the shaft of the armature being continued as the spindle of the screw. The cost of the launch complete was placed at $3,000. From that time onward, advances in electric launch work were made,

chiefly through the energy of Mr. Reckenzaun ; but the attention of the public was not arrested until a remarkable trip was made across the English Channel in 1887, when on September 13, Mr. Reckenzaun voyaged from Dover to Calais and back. His launch "Volta" (Fig. 7) was a boat 37 feet long, 6 feet 10 inches beam and 3½ feet draught, and was built of galvanized steel plate. Her battery consisted of 61 accumulators placed under the floor, and she was propelled by two Reckenzaun motors. The trip was made on a single charge of the batteries. The launch carried 7 passengers, and made a speed ranging between 6 and 12 miles an hour. The Channel was fortunately not as "choppy" as usual on this occasion, and the progress of the boat was so stealthy and quiet that one of the party captured with his hands a sleeping gannet afloat on the waves.

8. As a result of this very effective demonstration of the capabilities of electric power in navigation a great many boats were built in Europe, and their number has been steadily on the increase ever since. In England, the Immisch electric launch service was started, for example, in 1888, and its fleet to-day includes several very fine boats operated by the General Electric Power and Traction Co., Limited. The best known of these boats is the "Viscountess Bury," which is 65 feet long, by 10 feet beam, with a mean draught of 2 feet 9 inches. She has often accommodated 70 passengers, so that in this country she would carry at least 140. Her dining saloon will seat 24 passengers. She has a battery of 180 E. P. S. accumulators which can be arranged in two or four parallels, as desired. Her motor will easily develop 10 horse-power at 1,000 revolutions. It drives a two bladed screw of phosphor bronze, 19½ inches in diameter, with a pitch of 13 inches. Her speeds are 4½ and 6 miles an hour. Further mention will be made of this class of boats, belonging to fleets in regular service, in a later chapter.

9. The first storage battery launch put in commission in America was that owned by the brothers Anthony and Frederick Reckenzaun, called the "Magnet" (Fig. 8). She

was built by the latter, at Newark, N. J., in 1888, and was 28 feet long, with 6 feet beam and 3 feet depth amidships, and drew 2 feet 6 inches at the stern. Her Reckenzaun motor drove directly a gun metal two bladed screw of 18 inches diameter. The motor weighed 420 pounds and fitted snugly into the curved lines of the hull. The boat was equipped with 56 cells of the Electrical Accumulator Company's make, disposed under the floor along the keel, each cell weighing a little over 40 pounds and the total weight being about 2,400 pounds. The motor and cells were controlled by two switches placed near the pilot's seat in the stern. One of these started, stopped or reversed the motor. The other connected the battery in a series of 56 cells or grouped it in two parallel series of 28 cells each. The total weight of motor, cells, switches, wiring, screw, etc., was about 3,000 pounds. With the cells in parallel, the motor and screw made about 540 revolutions per minute, consuming an average current of 33 amperes, or nearly 2½ electrical horse-power and driving the boat at a speed of from 6 to 8 miles an hour. One charge was good for a 10 hours' run, covering a distance of from 60 to 75 miles. With the cells all in series, a speed of 10 to 12 miles could be made. But in this case both the electromotive force and the amperes were about doubled, and the net result was that while speed was gained, the duration of the discharge was cut down to about one-quarter of the normal time and the actual mileage to about one-half. Besides supplying current to the motor, the cells lit up seven 16 candle-power incandescent lamps and a 100 candle-power lamp placed in a reflector. The cells were charged from the factory plant of the Electrical Accumulator Co. at Newark, over about 1,500 feet of aerial line of No. 8 B. & S. wire, the current being from 20 to 30 amperes at from 140 to 150 volts.

This boat which would carry about 20 passengers sitting back to back along the centre under an awning, was sent some time ago to California, but not until after she had done a great deal of work on the Passaic River, Newark Bay, the Staten Island Kills, and even in New York Bay and the Hudson River. It was the good fortune of the present writer to make a long trip in her, in 1888, from New York to Newark, across the path of the ocean

steamers, the whole journey back and forth between the
two points representing from 50 to 60 miles at least.
Figs. 8a and 8b show the boat in plan and longitudinal
section.   She is still in use at San Francisco.

10. As in Europe, so here, the last few years have seen
a very general adoption of electric launches for private
and special use.   Reference will be made in a succeeding
chapter to the use of such launches at the World's Fair;
in the present chapter we propose simply to take note of
individual boats of importance.   One of the foremost
American patrons of this type of craft has been Mr. John

FIG. 8.—THE RECKENZAUN BOAT "MAGNET," ON THE
PASSAIC, N. J.

Jacob Astor, who has long made electricity a study and
whose personal interest in electrical navigation has given
the art a most valuable and helpful impetus in this country.   During 1893 he put into commission an electric
launch called the "Corcyra," illustrated (Fig. 9), page 18,
built in accordance with his designs, and those of Mr. J. C.
Chamberlain, whom he consulted.   She was so immediately
successful that her lines and general plan were adopted by
the American and Russian naval authorities.

While being entertained by Mr. Astor at Rhinebeck-on-
the-Hudson, the Grand Duke Alexander, of Russia, was
much pleased with this launch and learned that a similar

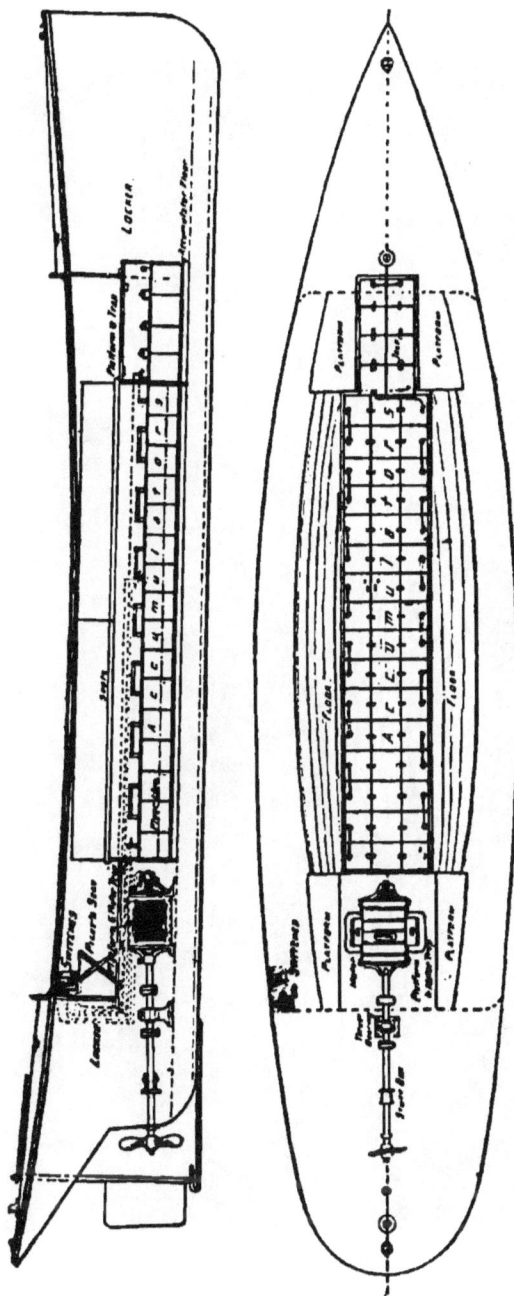

FIGS. 8a AND 8b.—THE RECKENZAUN ELECTRIC BOAT "MAGNET."

FIG. 9.—MR. J. J. ASTOR'S ELECTRIC LAUNCH "CORCYRA."

launch had just been completed by the General Electric Launch Company for the U. S. Government cruiser "New York," to be used as the Captain's gig. He shortly afterwards had an opportunity to inspect this electric gig also, and took such a fancy to it that, at the request of the Department of State, the Navy Department directed the contractors to deliver the launch to him and to begin the construction of a duplicate for the cruiser "New York."

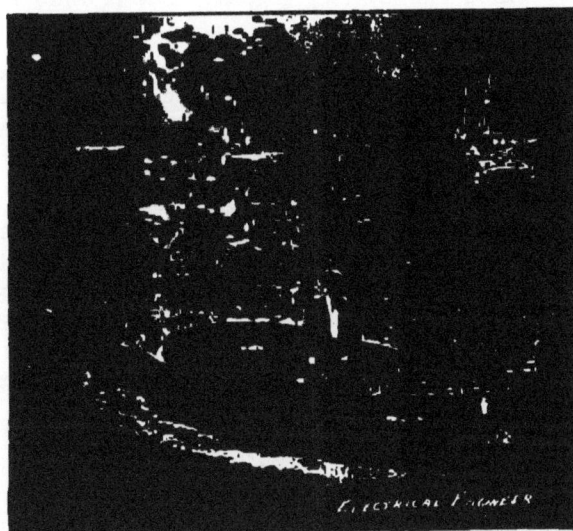

FIG. 10.—ELECTRIC GIG FOR THE GRAND DUKE ALEXANDER.

This electric gig, which is illustrated in the accompanying engraving (Fig. 10), is 30 feet long, 6 feet 8 inches beam and 22 inches draft. It is equipped with 66 storage batteries capable of delivering 3 horse-power to the motor for a period of 10 hours with one charge. The motor normally has a speed of 650 revolutions at 3 horse-power and is able to propel the boat at a speed of 6.4 miles per hour. The motor, however, is capable of developing 12 horse-power for a spurt of 5 or 10 minutes, increasing the speed of the boat to about 10 miles an hour. The batteries are placed underneath the seats and beneath the flooring, leaving the entire boat space for passengers. The motor is placed near the centre of the boat and underneath the flooring, being directly connected to the propeller shaft. The controller for regulating the speed is located near the wheel.

The launch will be used hereafter for the personal service of the Grand Duke in his cruises around the Russian bays and waters.

Mr. Astor has more recently had built for his personal use a twin screw 46-foot cruising launch called the "Progresso," which is shown in Fig. 11. She is undoubtedly

FIG. 11.—MR. J. J. ASTOR'S 46 FT. TWIN SCREW CABIN CRUISING LAUNCH "PROGRESSO."

the largest electric boat afloat in America, but it is understood that her owner has even more extensive plans under consideration. It is to be hoped that they may soon be realized, for the benefit of the art. The "Progresso" has twin propellers of 18 inch diameter, and each motor makes 800 revolutions when working up to capacity, viz., 4 horsepower each. She has 136 cells of battery, and makes from 8 to 12 miles an hour easily.

11. The Universal Electric Launch Company, with works at Nyack, N. Y., made recently an interesting test of a new 40-foot launch embodying several improvements. The boat is 40 feet long, 6 feet 6 inches in beam and normally draws 2 feet of water, giving a displacement of about three

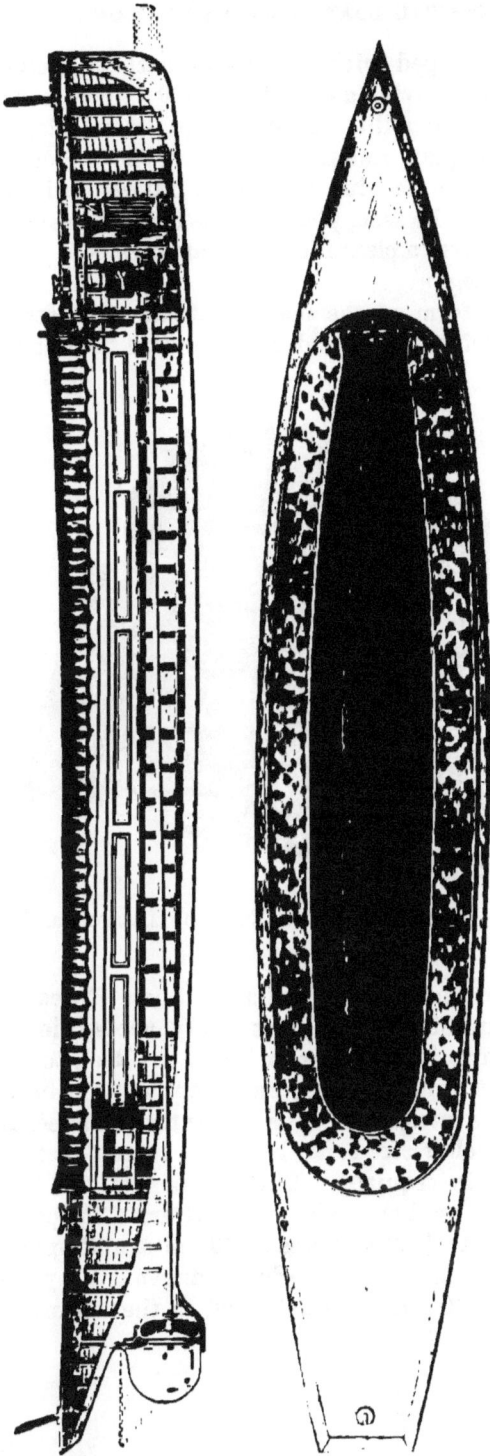

Figs. 12 and 12a.—The "Universal" Electric Launch, Equipped with Riker Motor.

tons. It is equipped with 72 cells of battery, weighing 3,000 pounds and having a capacity of 150 ampere hours.

The accompanying illustrations (Figs. 12 and 12a) show a sectional and plan view of the boat. The motor, a 10 horse-power Riker machine running at 600 revolutions per minute, is, as will be seen, placed in the bow. The enlarged view (Fig. 13) gives a clear idea of its arrangement and the method of setting.

The general form of the machine, that of a letter V, admits of its being placed under the forward deck in small

FIG. 13.—MOTOR OF RIKER ELECTRIC LAUNCHES.

boats or under the flooring in larger boats if desirable, where it is out of the way, yet easily accessible. The toothed armature is very low and near the keel, to which the machine is firmly bolted, and there is no vibration.

The cells are arranged to be instantly changed from series to parallel or *vice versa* by a pull switch, as normal or high speed is required. During the test the launch made 5½ miles an hour with the batteries in parallel, giving a current of 20 amperes at 70 volts, and spurts of eight miles an hour with the series arrangement, by which 43 amperes at 140 volts are delivered to the motor. The test of speed was considered very satisfactory, especially in view of the fact that the boat had been in the water all summer and the hull was consequently very dirty.

This launch acquitted herself most gallantly during the storm that raged along the Atlantic coast, August, 1893. She was at that time lying at anchor off Stamford, Conn., directly to leeward of a large naphtha launch, which, in spite of all the power of her engine running at full speed to take the strain from the cable, dragged her anchor and drifted down upon the Riker boat. In order to escape a

FIGS. 14 AND 14a.—PINNACE FOR STEAMSHIPS AND MEN-OF-WAR.

collision the latter was forced to slip her cable and rely upon her motive power alone. This manœuvre was successfully executed. The launch forced her way directly against the heavy sea and the gale, weathered a dangerous reef and then, turning in the trough of the sea, ran out of danger.

12. One more boat deserves inclusion in this group, namely, the electric pinnace shown in 1891, by the Electric Power and Traction Company at the Naval Exhibition in London; and here illustrated in Figs. 14 and 14a. It is

designed on the lines of the English Admiralty steam pinnaces and is 36 feet long, 7 feet beam, with a maximum draught of about 2 feet. It is built in pine with oak stem, stern, and stern posts, and is bright all over. The accumulators, 50 in number, are arranged in teak boxes under the seats, as shown in section, and are so placed as to be easily removable when necessary. The whole is so strongly built that the pinnace is slung on davits with all electrical equipment ready for use. The cells are charged either with the pinnace slung in the davits or moored alongside the ship. The motor is much more powerful than those usually supplied by the company for use in the launches on the Thames and at Windermere, and a speed of 11 miles per hour is obtained. Since these boats are principally used for shore purposes, speed and power have been considered of more importance than duration of run. For shore and harbor pinnaces electricity possesses many advantages over steam; and they will no doubt be largely in demand before long. One of these pinnaces, 40 feet long, 7 feet 9 inches beam, and 2 feet draught, with a handsome teak cabin, has been built to the order of the naval department of the Russian government.

CHAPTER III.

STORAGE LAUNCH FLEETS AND PASSENGER BOATS.

13. It is evident that storage battery launches offer many advantages for the purposes of miscellaneous work on a large scale, and that if a good charging station be provided there is little limit to the number of boats that can be operated from it. A peculiar feature of attractiveness is that the boat can be put in the hands of any intelligent person and that it is free from many if not most of the objections that attend the employment of manual labor, steam power, or sails; while it has some excellencies that even the convenient naphtha launch does not share. With an electric boat, there is an absolute freedom from noise, heat, smoke, gas, cinders, ashes, etc. There is nothing to explode; the current used is of a perfectly safe low pressure free from any risk of giving shock; and if the boat is properly wired there should be no danger whatever of fire. The boats are necessarily stable, the centre of gravity being low; the batteries are under the floor or seats and can be disposed as desired. This also increases the carrying capacity enormously, as the entire deck is free for passenger occupancy and there is no need for any movement of the crew, should the boat be manned. An element of economy is that this form of motive power dispenses with a trained crew or engineer, and that when the boat is needed, a simple movement of the switch renders part or all of her concealed power immediately available. As soon as the boat stops, all consumption of energy or fuel ceases.

14. These are some of the reasons that have led to the equipment of storage battery fleets, which, so far, have been of two kinds, namely, fleets all the boats in which are intended to perform a common duty in the nature of ferri-

age ; and, secondly, fleets any boat of which can be hired individually for the day, week, or season. Before passing on to describe both classes, it will be interesting to quote here some figures prepared in 1890 by Mr. Fred. Reckenzaun, at the request of the writer, to show how an investment in such a fleet might be made and what were the possibilities of return. The calculations and separate items may be taken as fairly approximate at the present time, subject to special conditions and to the steady lowering of the price of all electrical apparatus from year to year.

ESTIMATE OF COST

of a fleet of 12 electric launches, each 28 feet long, 6 feet beam, carrying one ton of storage batteries, to run 6 miles per hour for 60 miles with one charge :

12 hulls complete, with interior fittings (battery troughs, seats and lockers), fixed roofs, shades, flag staffs, steering wheels, etc., -	$6,600
12 tons storage batteries (cap. 16,240 watt hours per ton) at $560 per ton, - - - -	6,720
12 motors, at $400, - - - - - -	4,800
12 screw propellers, shafts, couplings, thrust bearings and stuffing boxes, - - -	1,200
Switches, wires, incandescent lamps (4 per boat), with fittings, - - - - -	480
Acid and labor of placing electric outfit, -	1,200
Seat cushions, ropes, boat hooks, tools, pumps, etc., - - - - - - - -	300

Total, 12 boats complete, in running order, $21,300 or $1,775 each.

CHARGING STATIONS.

Land and buildings (on suburban water front), say, - - - - - - - -	$4,000
Steam plant, 60 horse-power complete, erected,	4,000
Dynamo, capacity 40,000 watts, with accessories, erected, - - - - - - -	2,000
Charging circuits and appliances, erected, -	250
Mooring facilities, tools, etc., - - - -	500

Total cost of station, say, - - - $10,750
Grand total cost of 12 launches with charging facilities and real estate, - - - - 32,050

ESTIMATED COST OF OPERATION.

It is assumed that each of the 12 launches makes a daily run of 60 miles, divided into 6 trips of 10 miles each (3 round trips) during 5 months in the year:

12 pilots at $2.50 per day each, for 5 months,	$4,500.00
1 station engineer, at $3 per day for 5 months,	450.00
1 station fireman at $2 per day for 5 months,	300.00
1 station laborer at $1.75 per day for 5 months,	262.50
Coal (4 lbs. per horse-power hour, 60 horse-power for 7 hours daily), 112½ tons (for 5 months) at $4 per ton,	450.00
Oil, waste, miscellaneous supplies and incidentals for 5 months, say,	200.00
Labor, etc., putting boats in running order, at beginning and storing same at end of season, say,	360.00
Depreciation, per annum, on boats and propelling apparatus, at 10% on $21,300,	2,130.00
Depreciation of station machinery and appliances, at 6% per annum on $4,750,	285.00
Interest, per annum, at 6%, on invest. of $32,050,	1,923.00

Total operating expense, depreciation and interest,	$10,860.50
or $905.04 per boat per annum.	
Total mileage run per boat per month (60 per day),	1,800 miles.
Total mileage run per boat in 5 months,	9,000 "
Total mileage run, 12 boats, at 9,000 miles each,	108,000 "

Cost of operation, including running expenses, depreciation and interest, as per above estimate, = $10\frac{1}{10}$ cents per boat mile. The boats assumed can seat 20 passengers and over. If an average of only *one-half* of this number is constantly carried, paying fare at the rate of *one cent per mile* each, the receipts will equal the operating expense, depreciation and interest on investment, as above.

The boats, in this instance, run at intervals of about 17 or 18 minutes (allowing for short stops), 1½ miles apart, along the entire distance of 10 miles.

The cost and operating expense of electric launches will,

of course, vary with different sizes and speeds, which
the conditions and requirements of each distinct case
contribute to determine.

15. One of the earliest, largest and finest fleets for public
hire is that which was established in 1888, as already noted,
on the Thames, using Immisch motors, and is now carried
on by the General Electric Power and Traction Co., Limited, with headquarters at Platt's Eyot, Hampton. According to the latest information received, the Company now
possesses a fleet of 22 launches, 5 floating charging stations
and 4 permanent charging stations. The largest boat is the
"Viscountess Bury" (Fig. 15), to which reference has
already been made. She can be hired, it is stated, for twelve

FIG. 15.—THE "VISCOUNTESS BURY."

guineas ($60) per day, that sum including captain and assist-
ant, lock dues, etc. As she will accommodate 70 passengers,
it is evident that the charge per head is very small. The
"Omicron" is also 65 feet long, like the "Viscountess
Bury," but is only 7 feet beam. Then come a number of
boats ranging in size from 45 feet long to 35 feet long, and
another batch running up to 27 feet long, the charges for
hire ranging proportionately from $30 per day down to $15,
the sum named including wages for a deckhand, lock dues,
etc. These launches have long been a familiar sight on the
upper Thames, and are largely availed of by parties which
wish to enjoy its placid beauties at a minimum of cost,
fatigue or inconvenience. The passengers they have carried run up into the thousands. The boats ranging from 30

to 45 feet have motors weighing 350 pounds, and at full speed develop about 3 brake h. p. at 750 revolutions per minute. They are equipped with from 40 to 50 cells. The rules of the Thames Conservancy require a low rate of speed as a precaution against "wash," and the boats are therefore arranged to run at a normal rate of 6 miles an hour, or at a "half speed" of 4½ miles. At the higher speed they take 28 amperes at 85 volts and at the lower 21 amperes at 43 volts. The "Viscountess Bury" has 180 cells.

16. Another picturesque spot that has welcomed the noiseless, smooth-gliding electric launch has been Lake Windermere, with whose scenes the names of Wordsworth, Southey, Ruskin and others are so closely associated. It can easily be imagined that in such a place these launches were found preferable to steamboats. The charging station is driven by water power, an ideal source of energy in such surroundings. An old mill had been partially destroyed by fire, but as the water wheel mechanism remained intact, a Glasgow firm of boat builders conceived the happy idea of putting dynamos in the ruined building at Cockshott Point. The conditions were not such as to ensure a high efficiency for the plant, but water power is generally cheap, and a good rate of hire for the boats has been obtainable.

17. A boat that belongs in the category of passenger craft is the "Electric" transport, built for the British Government for the special purpose of conveying troops to and fro along the River Medway, between Chatham and Sheerness, two important military and naval depots. This boat is open, 48 feet 6 inches long, with a beam of 8 feet 6 inches, and built chiefly of mahogany and teak. She derives her power from 70 cells of 19 plates each. She has a draught of 2 feet 3 inches and will carry about 40 men.

18. A very interesting and important experiment in the utilization of storage launch fleets was made in Scotland, in 1890, at the Edinburgh International Exhibition, when the enterprise of the General Electric Power and Traction Co. led it to place a flotilla on the Union Canal. These boats

proved a great success, and an interesting account of them was read before the Royal Scottish Society of Arts, by Mr. A. R. Bennett, M.I.E.E., to whom we are indebted for some of the details.

These launches numbered four, and were built of steel, the hulls being designed by Morton & Williamson, of Glasgow. They measured 40 feet long over all, by 6 feet beam, 3 feet 1 inch from gunwale to keel, and drew 2 feet 1 inch when empty. Equipped with motor and cells, they weighed 3¼ tons out of the water. Each boat had a capacity of 40 passengers, with ample elbow room. They carried 50 cells of the E. P. S. boat type, each cell weighing 58 pounds complete. The cells were of ebonite, standing upon glass insulators filled with resin oil. These cells were ranged along the sides of each launch in two rows, 25 on each side ; and the lids, covered by cushions, served as seats. The motors were of a modified Immisch type and weighed 350 pounds each. The propellers were coupled direct to the motor shaft. A ball-bearing thrust block was attached to the motor bed and constructed in combination with the plain bearing of the motor. Fig. 16 shows the launch in longitudinal section ; Fig. 16a a midship section ; and Fig. 16b shows the motor and the ball bearings. The motor had four field coils with a resistance of .18 ohm when hot. These were in series with the drum armature, which had 48 coils and a hot resistance of .3 ohm. At

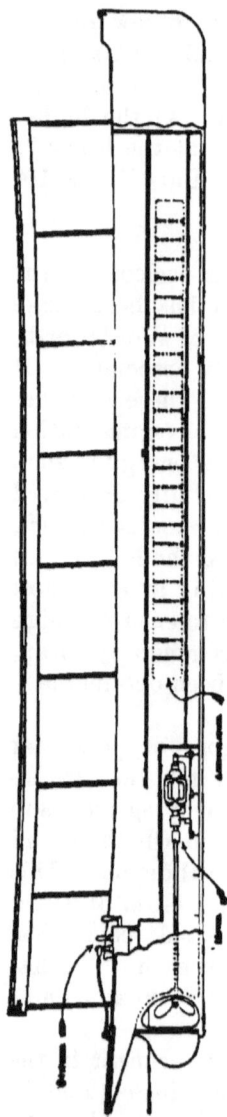

FIG. 16.—LONGITUDINAL SECTION, EDINBURGH EXHIBITION LAUNCHES.

its most favorable speed this motor had an efficiency of 85 per cent. When the launches traveled at the rate of 4½ miles an hour the motor efficiency was cut down to 75 per cent.

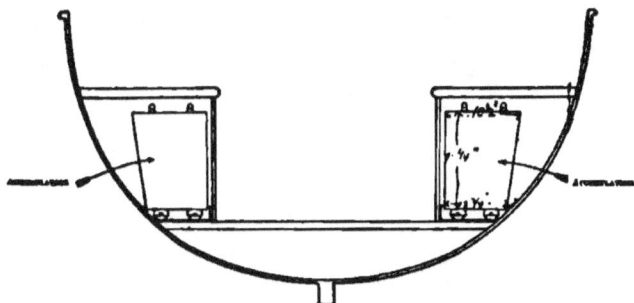

Fig. 16a.—Midship Section.

The controlling switch is shown in Fig. 16c. It had three levers.   The current from the accumulators was led to the switch through a lead fuse which opened at 42 amperes. For the first instant there was nothing but the ordinary conductor resistance of the coils to overcome, and a rush of current would have occurred but for the interposition of the series of resistances shown. The beginning of the movement admitted current through a resistance of 2 ohms,

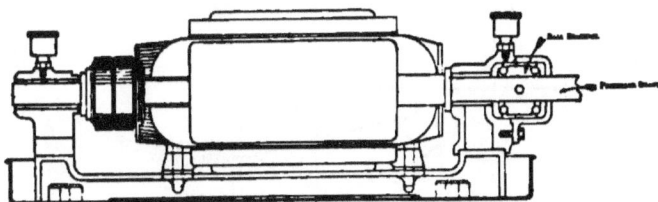

Fig. 16b.—Immisch Launch Motor.

which was gradually diminished as the lever passed the second and third contacts, until it reached the fourth, by which time the armature had started up and was developing its counter electromotive force, when full current was on. The second lever in the controlling switch set the

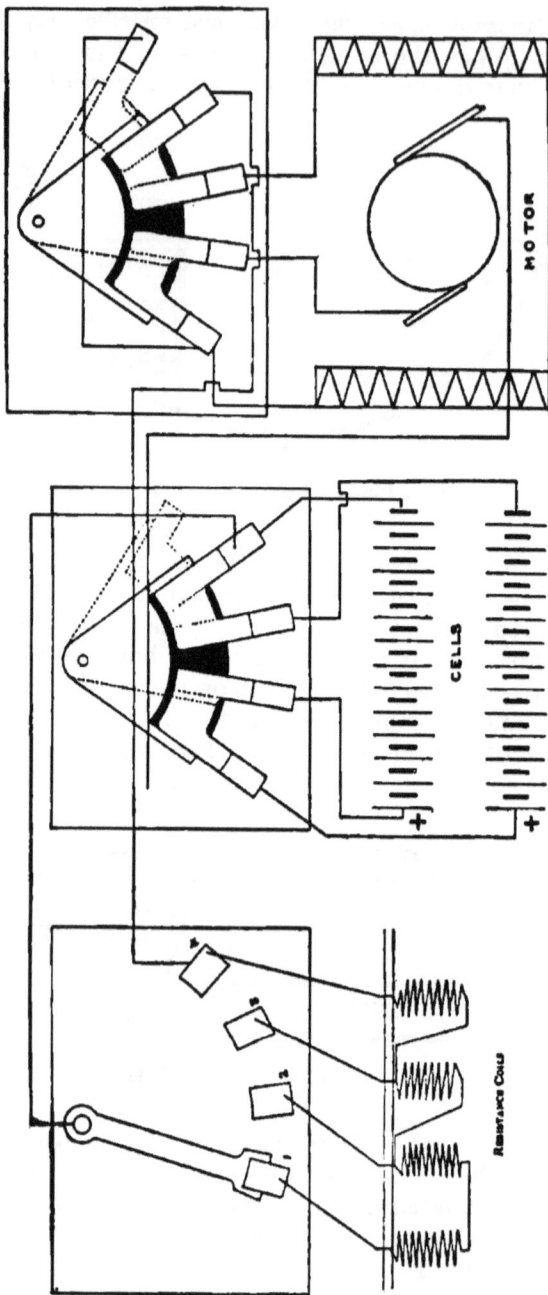

MOTOR

Reversing Switch.

CELLS

Series and Parallel Switch.

Resistance Coil

Resistance and Breaking Switch.

Fig. 16c.—Switches of Immisch Motors, Edinburgh Exhibition Boats.

speed at 6 or 4½ miles as desired by throwing the cells
either all in series or in two equal parallel groups of 25;
while the third lever was used for reversing, and simply
changed the direction of the current through the armature.
Neither the second nor the third lever could be shifted
while the current was flowing, the first lever locking the
other two mechanically while it was in the "on" position.
In this manner the motor was protected from harm, and the
current was guarded against waste.

FIG. 16d.—EDINBURGH CHARGING STATION AND ONE OF THE
LAUNCHES.

The propeller shaft was bolted directly to the armature
spindle, and passed through an ordinary water-tight gland
at the stern, the back and forward thrust being taken up at
the armature by means of the ball bearings already mentioned.   Propellers of various patterns were tried.   At a
speed of 4½ miles an hour, a three bladed propeller made
too much wash, and a phosphor bronze two bladed form
was adopted, having a diameter of 19½ inches and a pitch
of 14 inches.

19. These four launches were charged during the night, so that all of them were available for service throughout the day. The charging plant, Fig. 16d, consisted of an Immisch shunt-wound dynamo, running at 750 revolutions and delivering 120 amperes at 130 volts, or about 21 horse-

Fig. 17.—35 ft. World's Fair Launch, Making Circuit of Court of Honor.

power. The dynamo was driven by a 25 horse-power engine. For charging, the cells in each boat were put in series, and then the four boats were grouped in parallel. Each launch was charged with current sufficient to run it easily from 10 to 12 hours. The longest run made by any

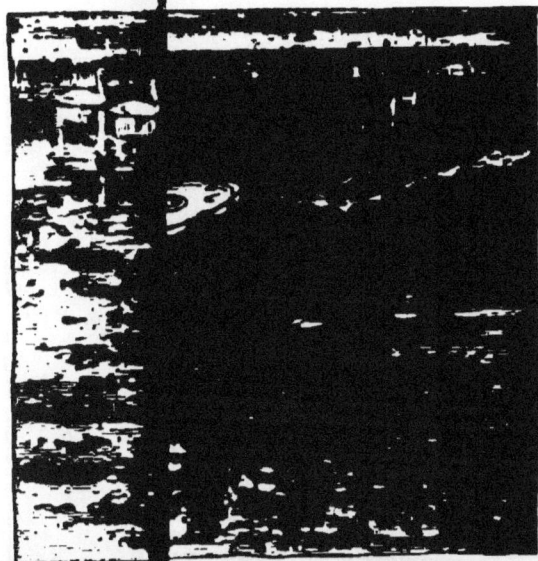

1
so
da
sh
del

po
en
se
E
ea

Fig. 18.—Electric l. a

boat was from the Exhibition to Linlithgow and back, a distance of 40 miles. The current was also used for signal bells and incandescent lamps.

The boats plied regularly along the Union Canal, and a four cent fare was charged. Between May 31 and October 11, the boats carried no fewer than 71,075 paying passengers, besides season ticket holders, officials and others entitled to ride free. On the busiest day, 2,560 passengers were carried.

20. It will be remembered that the World's Columbian Exposition in 1893, at Jackson Park, Chicago, was laid out by Mr. F. L. Olmsted so as to include a series of lagoons and canals, communicating with Lake Michigan and serving as a means of approach to nearly all the important buildings of the Fair. A course of about three miles was thus open to profitable use, and the competition for so valuable a concession was very active. Several concerns were bidders for the exclusive franchise, and each of them was required to build a sample boat to be submitted to trial tests in order to determine which was most suitable for the requirements of passenger traffic on the Exposition waterways. Steam, naphtha and electric launches were tested, and the privilege was awarded to the

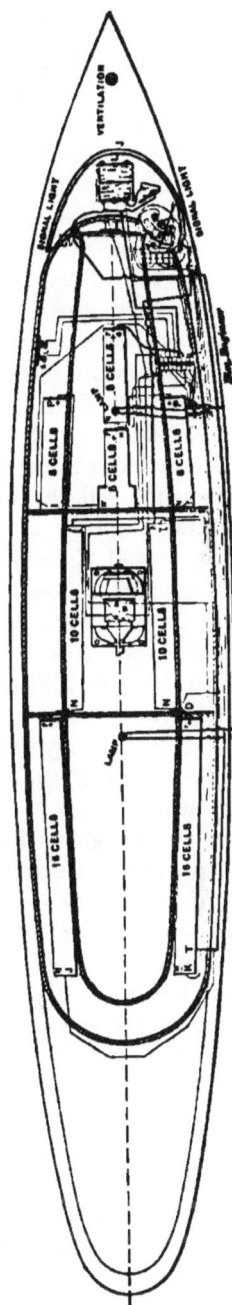

Fig. 18a.—Plan of Wiring, World's Fair Electric Launches.

Electric Launch & Navigation Company of New York, who thereupon equipped a fleet of 50 launches and placed them upon the waters of the Fair, while the Exposition equipped four boats for its own special use. The launches were fitted with Reckenzaun motors modified and constructed by the General Electric Company of New York, and with accumulators made by the Consolidated Electric Storage Company of New York. One of the boats is shown in Fig. 17, in perspective; in section in Fig. 18, which also shows the motor and its bearings.

Each boat was 35 feet 10 inches over all in length and 31 feet 6 inches on the water line. The beam was 6 feet 2¼ inches and the draught 27 inches. The hulls of these boats were constructed of white oak frames, with white cedar planking. The inner paneling, decks and other parts were mahogany. All the woodwork was finished in its natural color, thus giving a very rich appearance. The charging station was at the southeastern corner of the Agricultural building, in the South Pond. When the boats were to be charged they were laid up there, and whenever a boat needed repairs it was hauled up in its berth. The charging was all done at night, so as not to interfere with regular trips during the day.

The motor was protected by a box which rose flush with the main deck of the boat, and was so set that all working and wearing parts could be readily reached. The storage batteries were placed around the sides of the boat, under the seats and entirely out of sight. The motor was nominally of four horse-power, and was coupled direct to the propeller shaft, with a thrust ball bearing, in which the shaft ran. By this combination of direct coupling and thrust bearing all gearing and loss of power, as well as unnecessary noise and jar, were done away with.

The batteries used were of 150 ampere hours' capacity. Each boat had 66 cells, and these cells could be connected in three groups of 22 cells in series or in two groups of 33 cells in series. Fig. 18a shows the plan of wiring adopted for the boats, although in the diagram here presented 78 cells are enumerated. For the regular and moderate speeds, the batteries were grouped in three divisions, in series, with or without a resistance in circuit. The other grouping consisted of putting the cells in two groups in series, with or without resistance. The controller consisted of magnetically controlled switches, combined with mercury cups, operated by a lever at the pilot's right hand. Four speeds ahead and two astern were obtainable, and proved adequate for every purpose. A current of 18

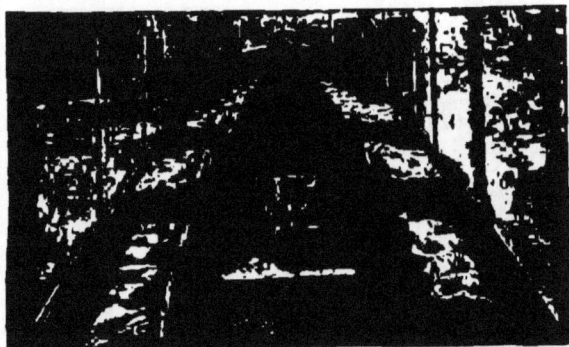

FIGS. 18b AND 18c.—ARRANGEMENT OF THE SEATS, BATTERIES, MOTOR, ETC.

amperes per group was generally used to charge after a run of 50 or 60 miles at nominal speed, and from six to seven hours was required. In case of necessity, however, a current of 30 amperes was used, when the batteries were charged in four or five hours.

Not one of the pilots or guards who managed the fleet

had ever before handled an electric launch, yet they experienced no trouble whatever from the first.   The control was by means of a small lever switch at one side of the

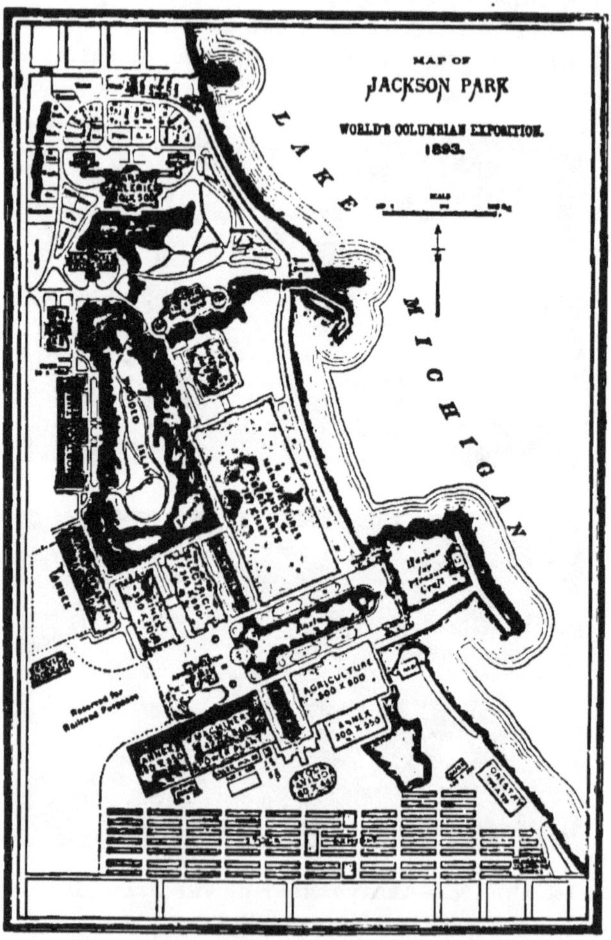

FIG. 18d.—MAP OF WORLD'S FAIR GROUNDS, INDICATING INTERIOR BASINS AND LAGOONS UPON WHICH THE ELECTRIC LAUNCHES WERE USED.

steering wheel, located in the forward part of the boat. This lever allowed of four speeds forward and two backward.   The nominal speed at which the boats were run was from six to seven miles an hour, but they had a reserve speed of from eight to ten miles. At the ordinary speed, the launches at the Exposition were in constant use from 12 to 14 hours a day on one charging, and the cost of this charging never exceeded 60 cents per day per boat. All the electrical parts except the switch were protected, so that there was no possible danger of shock, and there was no report at the Exposition of any mishap of this kind.

The greatest test the launches had during the period of the Exposition was on Chicago Day, when 622 trips, each trip of three miles, were made by the fifty boats. Six of these boats averaged 50 miles each, while 20 of them averaged over 40 miles, carrying on each trip about 40 people. There were no fewer than 25,000 passengers carried by the launches on that day, and entirely without accident. In fact, very few accidents occurred to any of the launches, and almost all that did happen were caused by the propeller becoming entangled in wires or other debris in the bottom of the lagoons, thus springing the shaft or bending the propeller blades. One bunch of wire was shown in the office of the company that was gathered up by a propeller and that would require a bushel basket to contain it.

21. During the month of October, 1893, the writer had frequent opportunities to watch the performance of these boats and found it most satisfactorily to the public and to all concerned. Believing that definite data would be valuable he secured from Mr. R. N. Chamberlain, the electrical engineer of the service, the following detailed figures up to that time :

APRIL 13 TO OCTOBER 1, 1893.

Mileage and Passenger Traffic.

Passenger trips of three miles each, - -	47,787
World's Fair special launch trips, - -	6,750
Special trips of regular launches, - - -	1,220
Trial trips, - - - - - - -	270
Experimental trips, - - - - - -	180
Total number of trips, - - - -	56,207

At three miles per trip, total miles,   -   - 168,621
Average miles per launch to October 1,   -   3,122
The total number of days the 54 electric
launches have been in service on the lagoons
of the World's Fair was,   -   -   -   -   6,594
Therefore the general average of miles per
launch per day was, -   -   -   -   -   25.57
Minimum miles per launch per day,   -   -   14
Maximum   "   "   ..   -   -   37½
Maximum miles, one launch. one day,   -   -   54
Total number of passengers carried from May 1
to October 1,   -   -   -   -   -   801,000
Maximum passengers carried in one day by
one launch,   -   -   -   -   -   -   464
Maximum number of people carried by one
launch for one round trip,   -   -   -   40

Operating Cost.

Average cost per launch per day for charging,
at 3 cents per electric horse-power,   -   -   55½c.
Average cost per launch per day for care and
repair of shafting, propellers, 54 motors, 162
packing boxes, 3,524 storage batteries, including labor for charging 54 controllers—
all the above being gone over every 24
hours,   -   -   -   -   -   -   -   43c.
Renewals of batteries per launch per day,   -   41c.
Renewals and repair material for all else per
launch per day,   -   -   -   -   -   9c.
                                                   _____
Total cost per launch per day, -   -   - $1.48½
Average cost per launch mile for labor and
material, exclusive of office expenses,   -   5½c.

The Fair lasted a month after October 1, and the number
of passengers reached about 1,000,000, before the service
was suspended. When it is understood that even at the
time these figures were compiled the launches had run on
an average four times the number of miles a launch owned
by a private individual would ordinarily cover during a
regular season, and that the operating expenses were

decidedly heavier than under ordinary circumstances, it will be acknowledged that the cost per launch mile is exceedingly low. Judging from the experience of six months with the 54 launches, Mr. Chamberlain believed that the expense can and will be, in the near future,

FIG. 19.—THE "VENEZIA"—35 FT. CABIN ELECTRIC LAUNCH UPON THE CANALS OF VENICE, ITALY.

brought down to as low a figure as three cents per electric launch mile where at least 30 launches are operated under like conditions.

The original cost of these launches complete was about $3,000. At the end of the Fair the fleet was placed on sale, and several of the boats have been bought for use in different parts of the country by persons who had witnessed their operation and had become desirous to introduce them on local waters for private pleasure or for public hire. The money-making ability of the fleet may be inferred from the fact that it had earned a total of $314,000, on the above showing, the round fare for the trip of three miles being 50 cents.

22. In competition with these launches at the Fair, were a number of gondolas, which were a graceful and pictur-esque element, but which were certainly not more poetic in ease and smoothness of motion than the electric boats. In fact the contrast in many practical features was so marked that a syndicate of Italians purchased one of the electric launches in September, 1893, and sent it direct to Venice to serve as the nucleus of a fleet on the historic canals. Under the fitting name of "Venezia," the little craft, which had been so busy at the Fair, made her reappearance in the waters so long associated with the song and oar of Adria's gondolier. It is needless to say that she created a sensation, for her quiet and graceful performance, so different from the noisy, fussy movements of the steam launch, gave immediate proof that in the electric boat the time-honored gondola had at last met a rival against which no halo of poetry and romance would help it greatly. Fig. 19 shows her at anchor opposite the Doge's Palace. She now has 68 cells in 4 groups, governed by a mechanical controller. In order to protect her hull against the weeds and action of the canal water, it has been sheathed with copper. The success of the boat has been indubitable, but it is understood that the fleet of which she is the proto-type will all be longer, so as to afford greater passenger capacity. Visitors to the World's Fair will note that her deck housing is slightly changed, it being now made to conform to the general style of the quaint cabin-top seen on the gondolas, in Venice, but not very much in evidence on the gondolas used at Chicago.

In the succeeding chapter reference is made to other launches and launch fleets that have been started, and details are given as to methods of operation in use or likely to be generally adopted for such service.

CHAPTER IV.

Special Features of Storage Launch Operation and Charging.

23. There are, all over this country, electric railways running through picturesque suburbs, watering places and summer resorts, skirting navigable streams or terminating at the shores of lakes and bays. Thousands of pleasure seekers and rest seekers are carried daily to these charming spots where the railway people turn them loose to fall into the hands of the boatmen for the rest of the day. One railway manager, however, conceived the idea of being his own boatman and keeping the public himself. He carried out his plan with whatever old apparatus he had at hand, or could secure; operated two boats all last summer and is so well satisfied with the success of his combination plant that this year he intends to enlarge his fleet materially and equip it with improved apparatus.

Fig. 22.—Charging Station.

The Seashore Electric Railway at Asbury Park, New Jersey, runs for some distance at the northern end of its route near the shore of Deal Lake, a very pretty little body of water separated from the ocean only by a low range of bare sand dunes and reaching inland in labyrinthine fashion, its bare banks changing with surprising rapidity to cool mossy terraces covered with trees and undergrowth. At the ocean end, the road makes a turn within about 50 yards of the lake. This point was selected for the "charging station," a view of which appears in Fig. 22. Before discussing this structure, however, it will be well to speak of the boats themselves, as they are unique among electric craft. The flagship, "Bonaventure," is a flat-bottomed boat 36 feet long, 7 feet in beam, draws 18 inches of water and will carry 36 people at the rate of 8 miles an hour. It is equipped with 110 cells of Julien battery in series, modified to suit existing conditions, as will be explained, and a Crocker-Wheeler motor of 2 horse-power running at a potential of 220 volts (Fig. 21). The motor shaft is 1 inch in diameter and drives an 18 inch propeller. The boat has run for 12 hours on a single charge. The "Dart" (Fig. 20) is 32 feet long and 7½ feet in beam with 2½ feet draught. Unlike the "Bonaventure," it is clinker built and has a round bottom. Fifty-five cells connected in series supply

FIG. 20.—CHARGING THE "DART."

current at 110 volts to a Crocker-Wheeler motor in the centre of the boat directly connected by a 1 inch shaft to a 22 inch two-bladed screw, making 600 revolutions a minute. The "Dart" carries 25 people at an average speed of 10 miles an hour and can make, it is said, 16 miles an hour at a spurt. A Crocker-Wheeler rheostat of enameled wire coiled about cylinders of asbestos paper to prevent vibration, controls the motor and is provided with stops for 7 speeds.

The batteries, as stated, are old Julien cells formerly used on street cars and, for some time before being assigned to their present work, occupied a conspicuous place in the scrap heap of the railway company. When it was decided to try them on the launches, each alternate plate was removed, increasing the distance between the remaining ones to three-eighths of an inch. This, of course, results in a much higher internal resistance, but, on the other hand, the cells do not give out quite so often, and the company preferred resistance, to investment, in new cells during the experimental stage.

FIG. 21.—CROCKER-WHEELER ELECTRIC LAUNCH MOTOR.

But the chief novelty is the method of charging. The street railway feeder, 50 yards away, is simply tapped and a line run down to what has been called the "charging station," but which is in reality a small box nailed to a post and containing a cut-out switch, an ammeter and a 50-ampere fuse (Figs. 20 and 22). The railway takes a current with a potential of 500 volts and this pressure is reduced for the launches by an amusingly simple process. After passing the switch, fuse and ammeter in the box, and the batteries in the boat, the circuit leads to an old iron pot sunk in the mud at the end of the pier. A section of iron pipe is stuck loosely in the dry sand above high water mark and this is connected to the railway return circuit. It will thus be seen that to vary the potential it is only necessary to push the iron pipe farther down into moist sand or pull it up a little where it is drier. The batteries are charged with a current of 15 amperes at about 140 volts.

Mr. A. S. Hickley, general manager of the Seashore Electric Railway Co., with whom the plan originated, is much pleased at the result of his first season's work, and, as stated, intends improving it this summer and putting it upon a really commercial footing. The fact that with such apparatus—all old except the motors—two old boats could be run for three months and make money without charging exorbitant fares, certainly shows that thought, work and capital expended in perfecting such a system can be well invested.

24. So crude a method as this is here described and illustrated chiefly with the idea of showing that the combination of storage launch work with electric railway work is a new and promising field, and deserves attention. At the present time many electric railway companies in this country are interested in suburban property and in rural pleasure grounds at the outer end of the roads. Several of the companies have gone so far as to improve existing pools and streams and have even laid out extensive sheets of artificial lake. Hence the opportunity already exists, but the chief difficulty of the work proposed appears to lie in the high voltage of the railway circuit—from 400 to 500 volts. It does not seem feasible to install in small launches for casual, or even for regular, use a number of cells sufficient to represent that voltage, even if two or three boats were charged with all the cells in series. But it is obvious that an easy way to meet the problem is to install a dynamotor, or motor-generator, at the charging point. By this means, the voltage of the railway circuit can be transformed down to that of the groups of cells, and a very successful use made of the railway current whenever it may be needed, either day or night. If desired, the motor-generator could be put on board a boat,

so that wherever it went it would obtain current from any railway circuit passing the shores of the waters it happened to be navigating.

Up to the present time, no demand for current has been of sufficient importance to make a special study of the question desirable in America, but the time has arrived when the increase in the number of electric launches depends on the frequency and accessibility of charging stations. A great many hotels along the sea shores, the rivers and the lakes have incandescent lighting plants, but as a general rule no arrangement has been made by any of them for running a tap circuit to a dock or landing where launches may be charged. Yet it would seem that money could be earned by the maintenance at most of these resorts, of one or two launches to be hired by the guests at a reasonable rate or for ferriage purposes; while in course of time the sale of current to visiting launches sailed by private owners might also be made a source of income. On the Thames, the General Electric Power and Traction Company has a tariff for the sale of its current to private launches. This tariff is based on a day's run and is estimated by the carrying capacity of the craft. Thus for a boat capable of carrying 30 passengers the charge has been about $4 (17 shillings); for 24 passengers, $3.75 (15 shillings); for 15, $3.25 (13 shillings and 6 pence); and for 10 passengers, $2.75 (11 shillings and 6 pence). These rates are fair and remunerative.

25. Leaving aside the isolated plants, it is safe to say that no large stream in this country near a town or city is far from a central station. As it is the business of any central to develop the sale of current, it is evident that the field of storage launch work is worthy the attention of local plant managers, either in the way of charging cells or in carrying leads to the water's edge, so that boats may come alongside to obtain supplies of current. This was begun last year in Boston by the local Edison Co., whose station backs on the Harbor. At Newport this year, the electric launches of Messrs. Astor and Vanderbilt will be charged from the local lighting station.

Where launches are attached to steam yachts or steamships it is easy to charge them hanging in the davits while the plant on board is running, and even to use the storage launch as a means of current supply during the day, if the demand is not large. If the launch is a big boat and no outside supply of current is obtainable, it should be possible to have a small charging plant on board, driven by an oil or naphtha engine, and to charge during the night or while at anchor. This arrangement is not complicated or expensive and might often prove advantageous. Mr. Astor has such a plan under consideration.

26. The Milwaukee, Wis., Electric Launch Company was formed recently with the intention of operating on the Milwaukee River, which is dammed at the city limits. Above that point, for a distance of about two miles, range a number of summer resorts, reached hitherto by steam launches. This new company has bought a fleet of electric launches to be run both on pleasure trips and as a regular passenger line, taking one hour to make the round trip. The electric street car line ends at the dam, and it is the intention, this summer, to charge the launches from the 500-volt circuit by putting three launches in series, with the object of avoiding any loss by resistance. The boats are already very successful there.

Another company which has entered the same field is the Altoona (Pa.) and Logan Valley Electric Railway Company, of Altoona, which has an artificial water of 13 acres in Lakemont Park, at the end of its road, and upon which it has placed a 25-foot General Electric launch. With this boat, passengers are carried for 10 cents, the round trip lasting 7 minutes. The current in this instance also is derived from the street railway circuit. A group of german silver coils has been hung upon poles in the open air, for the purpose, but it is understood that this is only temporary.

In Boston, the necessity of making better provision for the crowds seeking to obtain recreation and amusement on the waters in the public parks, has led to the acceptance by the Park Commissioners of a proposition under which a fleet of no fewer than 250 boats of all sizes, rigs and origin will be put afloat, in charge of Gen. C. H. Barney, who managed the electric fleet at the World's Fair. The boats will be distributed upon the waters of some eight parks, and will be either handled by attendants, or hired independently. It is intended to make electric launches a special feature of the work. As a matter of fact this is not the first time that public park authorities have taken up the subject, as a movement in this direction was made as long ago as 1892, when the Southport corporation, England, started a special carvel built electric launch, with remarkable success. This boat, "The Bonnie Southport," will seat 40 persons, and is 38 feet 5 inches long, 7 feet 6 inches beam, and draws only 2 feet 3 inches of water. She is driven by a 5 horse-power motor.

New Haven, Conn., is also the scene of the utilization of electric launches in connection with the trolley system, and similar work has been begun at Rochester and Buffalo. About three or four miles from New Haven is a very pretty sheet of water called Lake Saltonstall, to which the trolley road runs. Mr. G. H. Townsend, the owner, has begun the equipment of a fleet for the purpose of carrying passengers around the Lake, which is about 3¼ miles long. Twenty cents is charged for the round trip. The current is received by the batteries through a resistance, consisting of a tank of salt water in which two large iron plates are immersed. The resistance is regulated by varying the distance between the plates in the salt water solution, a handle being attached to them for the purpose.

An interesting experiment is, it is said, soon to be tried on the Gulf of Mexico, with a line of electric launches plying between a number of towns and ports along the coast, and carrying both passengers and freight. The boats required for such service are necessarily small, and hence owing to the heat of the climate, steamboats are not in favor. Several electric light stations exist along the coast and from one or more of these, the launches will obtain their supply of current. Moreover, the boats are easy to handle and the fact that no part of the deck space is occupied by machinery enables better provision to be made for passenger service, as well as allows larger space for a variety of small freight.

Another method of charging that has been resorted to is

that in use by the Earl of Aberdeen, Governor General of Canada, for his electric launch at Ottawa. The method employed there, very successfully, under the supervision of Mr. Osmond Higman, is that of taking the current from a 250 volt motor circuit, owned by the Standard Electric Co., a rheostat being used to reduce the current to the proper proportions. The charging current is 40 amperes under a pressure of 80 volts.

A novel use was made of an electric launch in 1890 by the London *Daily Graphic*, whose boat followed close behind the racing crews in the very fast Oxford and Cambridge boat race, and from which carrier pigeons, with sketches by special artists on board, were dispatched frequently while the struggle lasted.

CHAPTER V.

SPECIAL ELECTRICAL CRAFT—ROWBOATS, CATAMARANS, AND PADDLE WHEEL BOATS.

27. From various statements already made, it will have been evident that the range of applicability of electrical boats is by no means small, and that a number of very different types are possible. A few specific references to modifications or novelties will be helpful and instructive. The first example here shown is an electric rowboat, the design of Mr. J. C. Chamberlain. We have recently had the opportunity of testing her merits on the Harlem River, and are of the belief that in many respects she would prove very useful, wherever current is available. It is obvious that primary batteries can be used, but in this case the boat is fitted up with storage batteries, which are preferable for many reasons, and which will doubtless be used when possible. We show the boat in Figs. 23 and 23a, and cannot do better than quote figures supplied by Mr. Chamberlain as to dimensions, cost, etc. The boat illustrated has the No. 2 equipment, being 17 feet long and 46 inches beam.

Boat fitted with equipment	No.	1	2	3	4	5
Length	feet	16	17	18	19	21
Beam	inches	45	46	48	50	52
Draught	"	11	12	12	14	14
Freeboard with full load	"	8½	9	9½	10	11
Floating capacity (8 to 9-inch draught of hull)	pounds	800	950	1150	1400	1700
Speed—Miles per hour for ⎰ 3-4 hours		3	3½	3¾	4	4½
continuous run of ⎱ 6-7 "		2¾	3	3¼	3½	4
Speed rate for short spurt	miles	4	4½	4½	5	5½
No. of boxes of batteries		2	3	4	5	6
Type of motor and controller		No. ½	No. ½	No. 1	No. 1	No. 1
Charging current—volts		25	37	50	62	75
amperes		10	10	10	10	10
No. hours for full recharge		4-6	4-6	4-6	4-6	4-6
Cost per hour—cents		2	3	4	5	6
Total weight, boat with equipment		875	490	640	780	900
Price complete, boxed and crated, f. o. b. N. Y. City. Varnished boat		$400	$465	$570	$635	$705
Painted "		$390	455	560	685	695
Weight of equipment only	pounds	265	370	505	610	715
Price of same, boxed, f. o. b. New York City		$255	$315	$415	$475	$535

It may be added that each boat has regulating and reversing speed controller. She has no sail, but carries oars.
The boat is specially desirable as a tender to larger craft,
for hunting or fishing up shallow creeks or rivers, but is, of

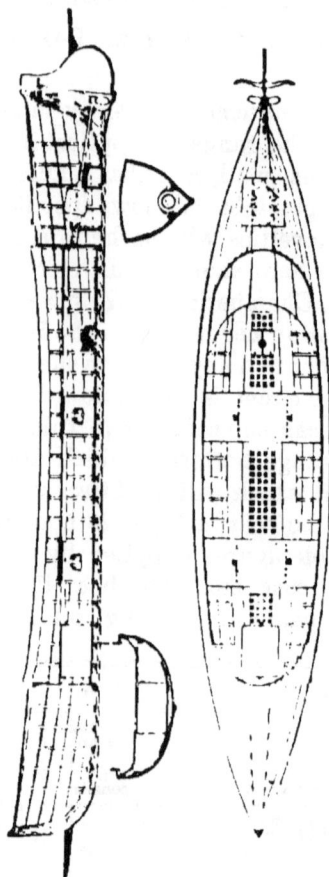

FIGS. 23 AND 23a.—ELECTRIC ROW-BOAT.

course, equally convenient for any casual use of business
or pleasure afloat, where it is deemed advisable to supplement, or diminish, the labor of rowing. As an adjunct for
fishing purposes, it is easy to drop overboard from this
boat, in circuit with the batteries, a little incandescent

lamp. It is well known that such a submerged light is an excellent lure for fish.

28. We may next consider an ingenious and novel variation in electrical boats, in the shape of a catamaran. The boat here shown (Fig. 24) was built and run very successfully in 1888 by Mr. Louis S. Clarke, of Pittsburgh, Pa., who sailed her upon the Conemaugh Lake that broke loose and so terribly devastated Johnstown lying below. This boat was built on the catamaran plan in the belief that less resistance might thus be offered to passage through the water, and that she might prove steadier than another boat of equal capacity of the ordinary build. The twin hulls

FIG. 24.—ELECTRIC CATAMARAN ON LAKE CONEMAUGH, PA.

were made of galvanized sheet iron. They were 22 feet long and 14 inches in diameter, and displaced about 2,600 pounds of water. A light platform 6 by 7 feet was constructed on the hulls, which were separated a distance of 4½ feet. On this platform stood the storage battery box, which served as a seat and contained 26 cells of storage battery weighing 400 pounds. They were charged from an 8-light dynamo on Mr. Clarke's steam launch. The motor to drive the catamaran was placed forward in a small box. The screw shaft stood at an angle of about 8 degrees from the water and ran clear to the stern. The little motor was of Mr. Clarke's own design, having a Gramme ring armature, developing about six-tenths of a horse-power and driving a 10-inch screw at about 450 revolutions per minute. The motions of starting, stopping, reversing and regulating the speed were governed by two small switches handily placed, and the steering was done by a small tiller in front of the seat. At night a little search light was used, or the craft could be lit up by one or two 50-volt incandescent lamps. The "Spark," as she was called, made a speed of about four miles an hour.

29. While it is not, perhaps, a distinct type of boat, the "paddlewheel" form deserves special mention, in spite of

FIG. 25.—TROUVÉ PADDLE-WHEEL BOAT.

the fact that so few examples are known. We illustrate here a "stern-wheeler" built by the ingenious M. Trouvé, who has supplied more than one such boat for use on rivers or waters where aquatic plants and weeds are abundant and might clog a screw. Fig. 25 is a perspective view of the boat, and Fig. 26 shows the method of mounting which enables the wheels to operate together or singly, and reversely.

Quite a novel piece of apparatus was constructed a few years ago at the works of the Freeman Electric Manufacturing Company, of New York City. It consisted of a boat mounted upon four wheels, and provided with an electrical equipment for its propulsion. The apparatus comprised a reciprocating motor which connected with the axle upon which the wheels were mounted. The motor itself weighed 87 pounds and the batteries 64 pounds, consisting of 20 storage cells. The vehicle, it was said, could be run at 10 miles an hour. Mrs. F. A. Truax, the designer of this apparatus, had in mind not only a vehicle for land travel, but one which also could be used for water. For this purpose the body was in the shape of a boat, and the wheels

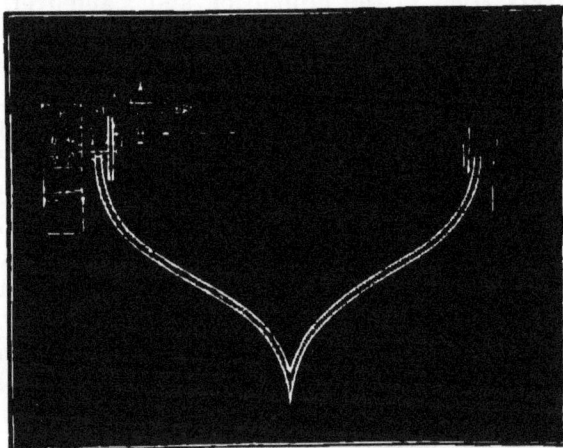

FIG. 26.—TROUVÉ METHOD OF CONNECTING UP THE PADDLE-WHEELS.

attached to the motor were also provided with paddles. Thus, upon entering the water, the apparatus was at once transformed into a boat with paddle-wheels. The woodwork of the boat was remarkably light, weighing about 33 pounds. The wheels weighed 24 pounds each, and the boat itself, with a load of 1,300 pounds, drew but 5 inches of water. No records as to the performance of this curious craft have been obtainable.

CHAPTER VI.

SUBMARINE ELECTRIC TORPEDO BOATS.

30. We have thus far been dealing with boats to navigate the surface of the water; but electricity has also been extensively tried in the propulsion and manœuvering of submergible boats intended chiefly, if not, indeed, wholly, for purposes of warfare. These boats constitute a distinct and important class. Without going into the history of submarine boats in general, or touching on the work of Van Drebble, Bushnell and Fulton, it may be stated that in 1887 the U. S. Navy issued a circular showing the requirements to be broadly fulfilled in the design and trial of a proposed steel submarine torpedo boat for the American navy. Some plans were actually submitted; but, so far as is known, the navy still remains without any such boat, whether electrical or otherwise. The main features in the stipulated requirements were that the boat should be able to make a speed of at least 15 knots an hour when running on the surface; 12 knots when "covered" by 3 feet of water; and 8 knots when submerged and offering no view of the object of attack other than one through water. She was also to be able to run for about 30 hours at full power, on the surface or covered, without detriment to her power for use under water; and when submerged she was to be capable of running at least 2 hours at 8 knots mean speed. If intended for "covered" and "submerged" work only, without using air draught, she was to be capable of running in that condition about 30 hours at full power. She was to be able to turn in a circle of a diameter not greater than four times her length, without reversing engines, and was to be able to pass from the surface to the plane below in 30 seconds. The shell of the boat was to be strong enough to withstand an external water pressure due to a submergence of at least 150 feet.

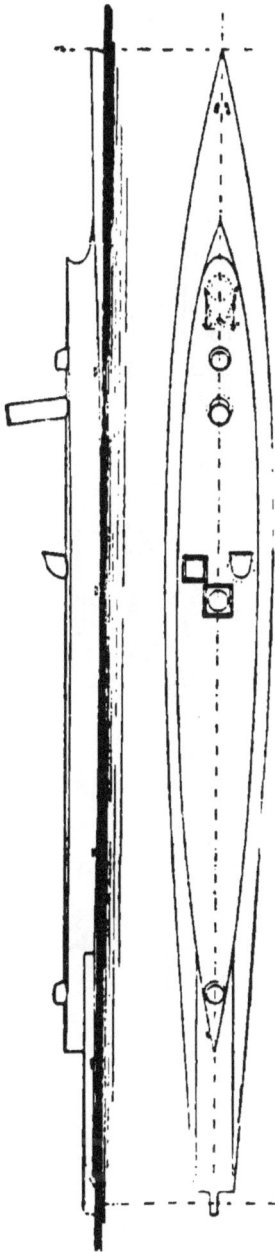

FIG. 27.—HOVGAARD DIVING ELECTRIC BOAT.

Against the bottom of a ship, running at speed, she was to deliver with reasonable certainty torpedoes carrying charges equal in minimum effect to 100 pounds of gun cotton. Besides this, she was to have means of obtaining an all-round view; of purifying the air for the crew so as to allow of at least 12 hours' submersion; of keeping the temperature within the boat down to 100 degrees Fah.; and of getting away from obstructions, lighting the interior, etc. The Navy Department limited the maximum displacement to 200 tons, when the vessel was submerged, but thought that about 90 tons would give the best results.

Before passing on to note what has been done abroad, in this field of work, mention is in order of the paper read by Lieut. Hovgaard, of the Royal Danish Navy, in 1888, before the British Institution of Naval Architects, on the "diving boat" for which bids had thus been invited by the American government. His boat then described is shown in Figs. 27 and 28. She has a fish shaped hull, 122 feet long and 12 feet beam, with a low superstructure surmounted by two conning towers, 1½ feet high. Her displacement at

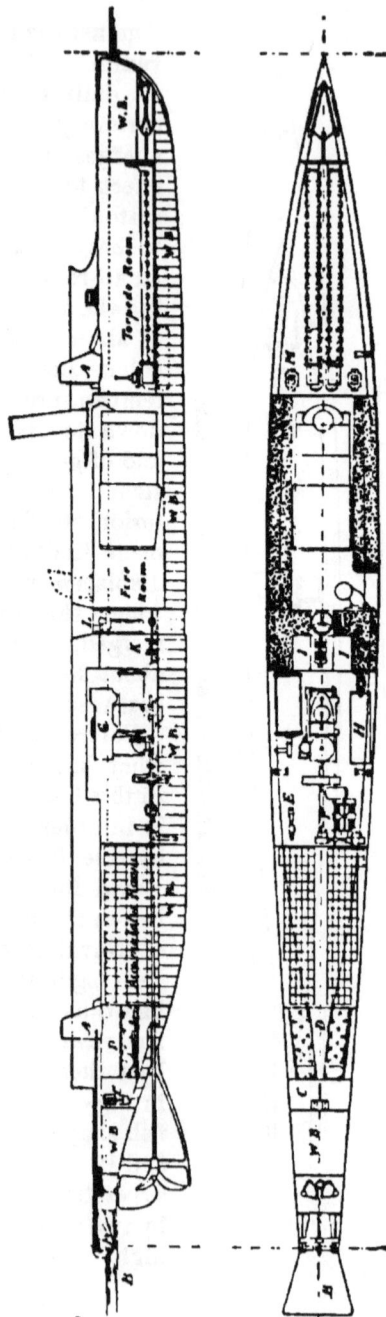

FIGS. 28 AND 29.—HOVGAARD DIVING ELECTRIC BOAT.

light draught was intended to be 171 tons; and, submerged, 196 tons. The ordinary power of the boat was derivable from vertical compound, surface condensing engines capable of developing 600 i. ii. p.; but during submersion she was to be driven by an electric motor developing 35 i. ii. p. taking current from a storage battery capable of giving a speed of nearly 6 knots for five consecutive hours. Of storage cells she was to carry no fewer than 540, of the lead grid type, each weighing 79 pounds and capable of delivering 35 amperes for 6 hours, the average electromotive force of each being 2 volts. The major portion of these cells—490—were to be placed in a separate battery room. All movements up and down were to be brought about by a small propeller placed near the centre of buoyancy in a vertical well going right through the vessel, and driven by a 5 horse-power motor also operated by storage battery current. Provision was also made for electric lighting, and for pumping out the water ballast by electric motor in about 30 minutes. The shafting of the propeller was so arranged as to shift over, when needed, from the steam engines to the electric motor.

31. Turning to what has actually been attempted in Europe, the first electrical boat we will describe is that designed and experimented with in 1888, and since, by Mr. J. F. Waddington, of Seacombe, near Liverpool, England.

The length of this vessel (shown in Figs. 29 and 30) is 37 feet over all, and her diameter amidships is 6 feet, tapering in a curve at each end. On the top of the vessel is the conning tower A, provided with ports at the sides and a water tight scuttle ǂ on top for entering the interior. The vessel is divided by the bulkheads BB into three compartments, of which the two end ones CC are used for storing the supply of compressed air, while the centre one is occupied by the driving, pumping and steering apparatus. The electricity used for working the motor is stored in 45 large accumulator cells D, which have a capacity of 660 ampere hours each, and are connected in series to an electric motor E, which drives the propeller F direct at about 750 revolutions per minute. The motor when working at full speed uses a current of about 66 amperes and 90 volts,

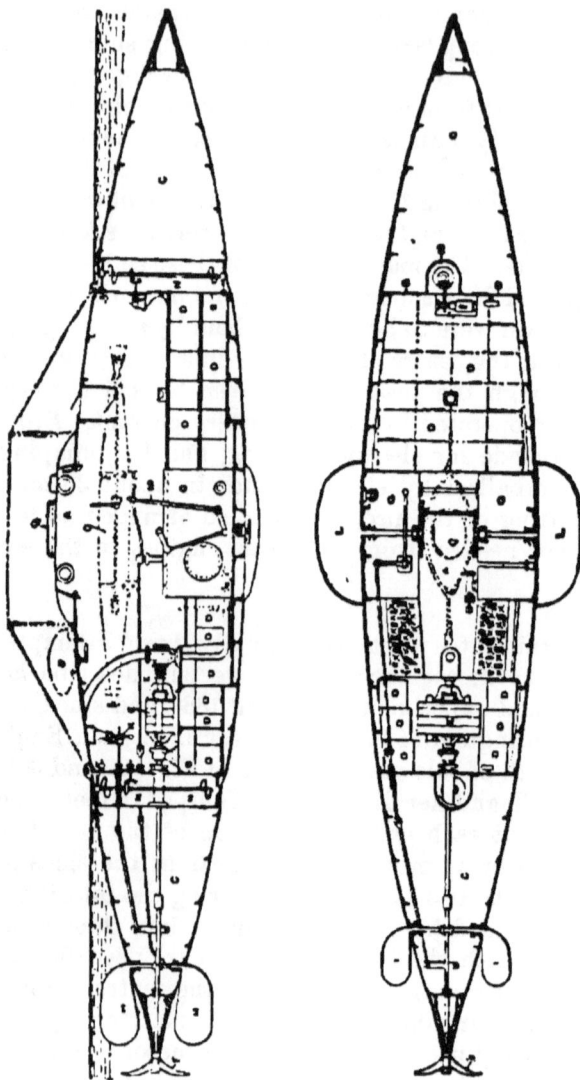

FIGS. 30 AND 31.—WADDINGTON ELECTRIC SUBMARINE BOAT.

developing 7.96 electrical horse-power, which would drive the vessel for 10 hours at a rate of eight miles an hour, thereby enabling the vessel to go eighty miles at full speed without replenishing the accumulators, while at half speed she would travel 110, and at slow speed 150 miles. The great advantages the inventor claimed in using electricity are that when once charged the vessel is ready to start at a moment's notice; that when stopped, the motive power is not subject to waste, as in the case of steam power; and that no heat or poisonous gases are given off to pollute the air in the vessel, which is an important feature in submarine vessels; also that a small sized class of these vessels can be carried on a man-of-war's davits, ready for instant use, the same as any ordinary launch. All the levers for manœuvering the boat are arranged amidships, within easy reach of the steersman, so that he can control the movements of the vessel, while looking out of the conning tower. A second man accompanies the steersman, to assist in case of emergency, so that the crew of the boat consists of two men only. On either side of the vessel large water ballast tanks G are provided, by filling which the buoyancy of the vessel may be reduced by several hundredweight when preparing to dive. At the after end are arranged two vertical rudders HHH and two horizontal ones II, the latter ones being used for keeping the vessel on an "even keel" when moving under water. It was found necessary to make the rudders II act automatically, as a man might lose his presence of mind when most required. The inventor at first provided a pendulum suspended from the top of the vessel and connected to the rudders II; but it did not answer readily enough to the action of the vessel, and he has since adopted an electric motor K working into worm gearing, and arranged in such a manner that the slightest cant up or down causes the rudders to be put over to meet it, and then stop in the same manner as a steam steering gear. At each side of the vessel balanced side planes LL are provided, which may be inclined at different angles by the lever M. Close to the bulkheads at each end of the compartments CC a vertical tube, open at each end, extends right through the vessel so as to leave a clear course for the water from the vertical propellers NN, which

are worked by separate motors, and may be used separately
or together, as the occasion may require. At the bottom
of the vessel a heavy weight is attached in such a manner
that it can be released from the inside in case of an emergency, such as a sudden leak, etc. The mode of working
the vessel is as follows: The water tanks GG are filled so
that only the conning tower is above water, representing only several hundredweights of buoyancy, the motor
started, and a speed of say four to five miles an hour
attained. Then the side planes incline by means of the lever
M, and the vessel is thereby submerged. By regulating
the angle of the side planes and the speed, any reasonable
depth may be maintained. If it be desirable to submerge
the vessel without forward motion, only the vertical propellers are used, and any reasonable depth may be maintained by regulating their speed. These vertical propellers
may also be used for regulating the depth when the vessel
is moving onward, but the side planes are considered preferable in that case. The fresh air is drawn from the two
end compartments cc, and the bad air escapes through a
valve as soon as the pressure inside becomes greater than
that outside. But the compressed air need only be used
when staying below a lengthened period, as the centre
compartment contains air sufficient for two men for six
hours without drawing upon the supply of compressed air.
The vessel carries two automobile torpedoes o, one on each
side, secured by grips, which can be opened from the inside of the vessel. By releasing these grips the propelling
motor of the torpedo is started, and it shoots ahead of the
submarine vessel. A mine torpedo P, is also provided for
attacking any vessel at anchor with the torpedo nettings
down, and which can be fired from a distance by an electric wire paid out from the submarine vessel. The dotted
lines indicate portable guard rails and stanchions, which
are put in place when the boat is above water—a very
necessary precaution in a seaway.

32. Considerable notoriety was achieved in 1888 by the
electric submarine boat "Peral," named after her inventor,
Lieut. Peral of the Spanish navy. She is shown in Figs.
32 and 33. Her lines resembles those of the famous White-

FIG. 32.—THE ELECTRIC SUBMARINE BOAT "PERAL" READY FOR LAUNCHING.

head torpedo, but with an ogival cross-section. Her length was 72 feet; maximum diameter 9 feet 6 inches; displacement, 86 tons; draught on water when operating at the surface, 2 feet 11 inches. She was equipped with two propellers, and with an Immisch electric motor driving each one; while the pumps were also worked by electric motor. She carried 613 cells of Julien storage battery, of which 125 were allotted to each propeller motor and 100 to the pump motors. The other cells were utilized for a projector, for inside lighting and for various other features about which the inventor and the Spanish officials were very reticent. Her nominal speed at the surface was 11 knots, and when submerged 10¼ knots. Besides her two screws for propulsion, the "Peral" had two others for immersion. If any accident occurred to the motors and the immersion screws were stopped, the boat rose at once to the surface without further aid, though, of course, the emptying of the water compartments hastened the ascension. To avoid an undue expenditure of power in connection with the immersion screws, the water compartments were filled to an amount which enabled a very slight motion of the screws to sink the boat to the required depth and maintain it there. The automatic apparatus which regulated the depth at which the boat was to work was designed on a principle somewhat similar to that of the aneroid barometer. A curved tube of elliptical section was placed in connection with the sea, and its deformations due to the alterations of pressure actuated a switch, by which the strength of the current going to the immersion screws was varied. The positions of the contacts of the switch were altered to suit the particular depth at which it is required to work. A very sensitive automatic electrical device was also employed to keep the vessel in a horizontal position. The apparatus consisted of a pendulum playing between two contacts. If the boat was not perfectly level from stem to stern, the pendulum touched one of the two contacts, and the result was that the corresponding vertical screw was actuated and the boat was righted. Sundry trials were made with this boat, but the silence in which she is now submerged would indicate that she was not an ultimate success.

FIG. 33.—THE "PERAL" PREPARATORY TO SUBMERSION.

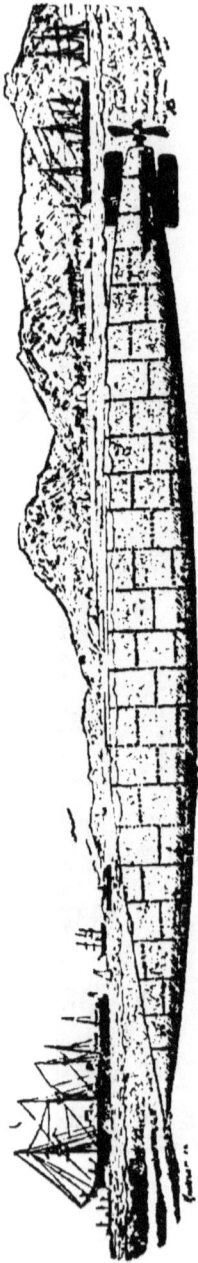

Fig. 35.—The "Gymnote" Submarine Electric Boat—Exterior View of Hull.

33. Another electric submarine boat of considerable fame is the "Gymnote" (Figs. 34 and 35), which was designed by M. Zédé, the engineer-in-chief of the Forges et Chantiers de la Méditerranée, whose plans were accepted for the French government by Admiral Aube. She was launched in September, 1888. Her shape is that of the Whitehead torpedo. Her length is 59 feet, greatest diameter 5 feet 11 inches ; displacement 29.5 tons. She has two horizontal rudders worked by hydrostatic pressure or at will, and two vertical rudders operated by the usual appliances. The motive power is electricity. The motor is of the multipolar type, designed by Capt. Krebs, well known from his work in aerial navigation, and weighs 4,400 pounds. It drives a 4-blade propeller, 4 feet 10 inches in diameter, connected directly to the armature shaft, and revolving at 200 turns per minute. The estimated speed is 10 knots for six hours, current being taken from a battery of 564 Desmazures alkaline accumulators, each cell weighing 38.5 pounds ; or a total weight of 9.66 tons. The boat is also steered and lighted by current from the cells. On one trial made in 1888, the boat dived 23 feet and traveled 1,600 feet at a speed of about 4 knots per hour, everything working well, although certain improvements at once suggested themselves.

The French Minister of Marine reported in 1891 that the "Gymnote" had "solved in a satisfactory manner the problem of submarine navigation, that is to say, the running

Fig. 34.—The Submarine Boat "Gymnote" at Work.

of a direct course under water, toward any determined point. This was proved last year at Toulon, when the 'Gymnote' ran outside of the harbor and directed her course under the water towards the 'Couronne' without deviating from the straight line for an instant." The Minister remarked, however, that such a boat must have weapons to fight with, and that these, with improvements,

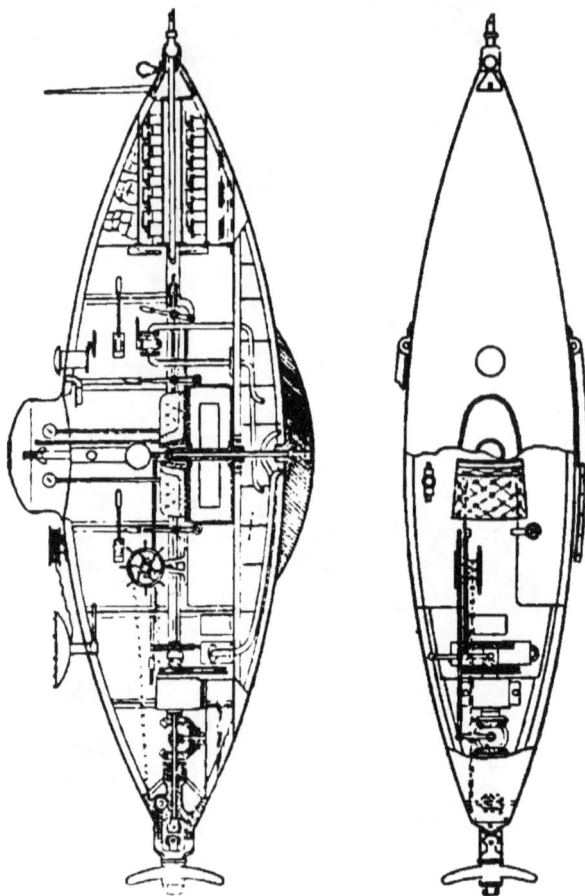

FIGS. 36 AND 37.—THE "GOUBET" ELECTRIC BOAT.

were to be furnished and embodied in a new boat called the "Gustave Zédé," to cost $223,800.

34. Another French submarine boat, of which it is said that the Russian navy has adopted it "provisionally," is the "Goubet" (Figs. 36 and 37), built on the lines of the

ordinary fish torpedo.   She is about 18 feet long, 5 feet in diameter, and has a total weight of 7 tons.   Her hull is of cannon bronze.   During the first trial made of her at Cherbourg, in 1889, to test her system of aeration, the two men who constituted her crew were hermetically sealed in the boat for 8 hours at a depth of 33 feet.   At the end of the experiment, the men when they came up to get some coffee, after their dinner, were perfectly comfortable and it was found that there remained sufficient oxygen to have prolonged the experiment for 25 hours.   The oxygen supplied for breathing was compressed at a pressure of 70 atmospheres in steel tubes 3.93 feet long, 4.72 inches in diameter, weighing 66 pounds.   Minor advantages alleged to exist in the case of the "Goubet" are the possible substitution or supplementing of the electric power by "submarine oars;" the provision by means of a heavy mass of lead attached to the keel and detachable at the touch of a button, for the instant coming to the surface, in the case of any accident; and the ease of steering, which is done by means of the driving propeller.   The report of the French authorities has not been at all favorable as to the capabilities of this boat for actual work.

35. In 1888, the French government commissioned the Compagnie Générale des Bâteaux Parisiens to build a submarine boat to be used in destroying submarine mines. Her greatest length is stated at 14.9 feet; her diameter at 5 feet 4 inches; her complement, two men.   She is driven by an electric motor taking current from a primary battery and is lighted by electricity.   No reports of her performance have fallen under the writer's eye, and this remark applies equally to some of the submarine boats built by Russia.   There is no desire on the part of the various governments to make public their tests in this direction, whether success or failure attend them; and hence the difficulty in furnishing even the simplest details of construction.   It is evident, however, that any great naval war would bring to light what has been done secretly in this field.

36. One of the most interesting submarine electrical boats is that designed by Mr. George C. Baker, of Chicago,

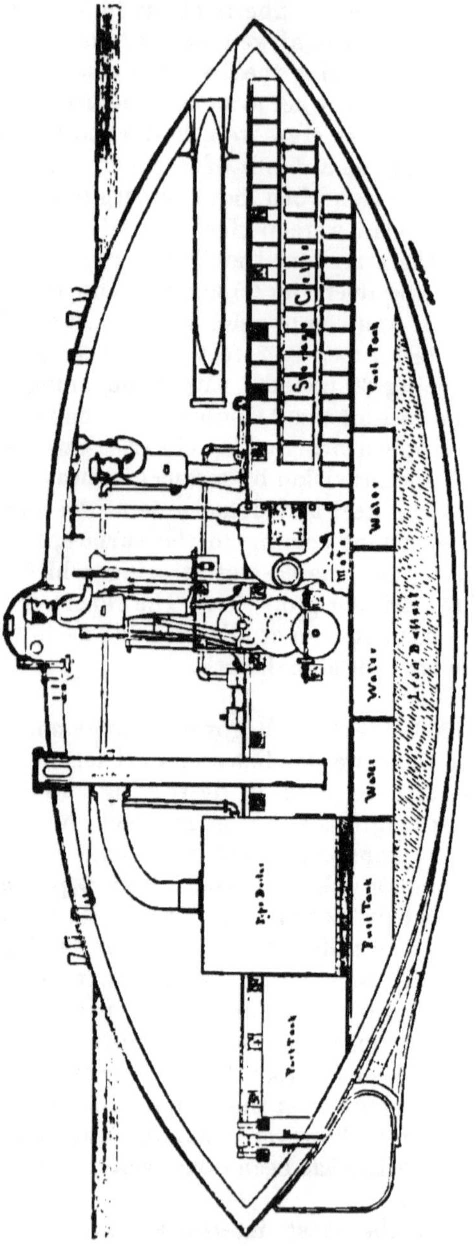

FIG. 38.—THE BAKER SUBMARINE ELECTRICAL BOAT.

and tried at Detroit, Mich., in 1892. She is shown in section in Fig. 38, and afloat in Fig. 89. She is built of wood, and has about 75 tons displacement, divided as follows :—Hull, 20 tons; ballast, 30 tons; storage battery cells, 10 tons; engine, boiler and gearing, 8 tons; and motor, 3 tons; leaving 4 tons buoyancy. At normal draught, about 2 feet of the crown of the hull remains above water. The shell proper is six inches thick with a sheathing an inch thick. It is built of strips of 3-inch oak, 6 inches wide, nailed flatside together, with 9-inch spikes. The boat is braced horizontally across its centre line by eight

FIG. 39.—APPEARANCE OF BAKER BOAT WHEN READY FOR
GOING UNDER WATER.

6 by 6-inch oak beams. The interior of the shell at the centre is 13 feet deep by 8 feet wide, and the outside length of the boat is 40 feet. The arrangement of the parts and mechanism is shown in the diagram. It is the plan of the designer to employ the steam plant to drive the dynamo and thus charge the storage battery. When the charging has taken place at any convenient point, the smokestack is drawn down into the boat and the fire is extinguished by closing the air-tight furnace doors. The cells may then be turned on again to the circuit, but this time to run the dynamo reversed, as a motor. The cap of the smokestack is a valve, which when the stack has been pulled in closes the only opening to the boat other than the manhole.

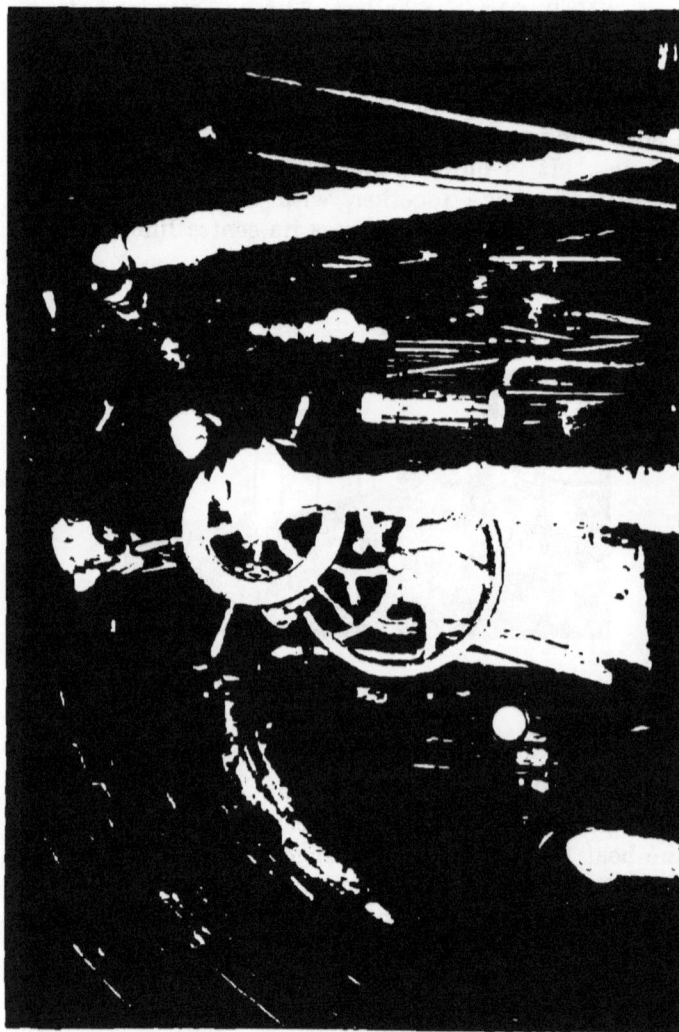

FIG. 39a.—INTERIOR OF BAKER BOAT WHEN IN OPERATION UNDER WATER.

The pilot stands on an elevated platform in the centre of the boat, his head being surrounded by the dome turret with five plate glass windows, four at the sides and one on top. Entrance and exit are made through the turret as a manhole. The turret cap is provided on its under edge with a gasket of rubber tubing. To seal the vessel, the lid has only to be swung around on the stud at the side, which forms its pivot or hinge; and then by turning a nut on the lower and inner end of the stud, it is let down and screwed tightly into place.

The side screws are so geared as to regulate the depth of submersion as well as to propel the boat. Mr. Baker's theory in regard to such boats has been that a boat should be forced under water by her screws rather than be sunk entirely by means of an added weight, such as water drawn into reservoirs on board. In other words, by the introduction of water into reservoirs, in addition to ballast already on board, the vessel's buoyancy could, at will, be reduced to a minimum; and when thus suspended or balanced in the water, the screws can be run at the proper angle to force the boat under. By driving the screws on an angle, the boat could thus be navigated at any reasonable desired depth. Another theory relates to the proper location of the propellers. Mr. Baker's belief is that the best plan is to employ two screws, each so arranged that it can be set from within the boat to revolve at any angle in a plane parallel with the vertical centre plane of the boat. These screws are therefore connected so that they may propel the boat backward or forward, on the surface or below, force it down for complete submersion or bring it to the surface. By placing the propellers also at the point of the boat's centre of gravity, Mr. Baker has sought to secure greater stability and to maintain the craft, under all circumstances, with its keel parallel to the surface of the water.

The equipment of the boat deserves brief mention in detail. The Willard engine is of the marine type, reversible, with links, etc., of nominal 35 horse-power. The Roberts boiler is rated at 60 horse-power, of the ordinary marine pipe type, and tested up to 220 pounds pressure. In connection therewith is a Worthington duplex pump

FIG. 30b.—BAKER BOAT AFTER SUBMERSION.

4½ by 2½ by 4 inches, employed to feed the boiler and to draw the water ballast out of the reservoirs shown in the keel. No filling of the reservoirs by pump is necessary, as the water runs in of its own accord.

The electrical plant comprises a 50 horse-power Jenney motor and 232 Woodward storage cells. The motor is built for an electromotive force of 200 volts and runs at a maximum speed of 900 revolutions. It is geared to the two propelling screws, which are four bladed, and 24 inches in diameter from tip to tip of blade ; and the screws run at 300 revolutions, which gives an estimated speed of about 8 or 9 miles an hour. When run by the engine as a generator of current, the dynamo works at 1,025 revolutions per minute, and at a charging pressure of 220 volts. The cells are grouped in four sets of 58 each, and are discharged in two sets of 116 cells each. At the top of the boat is a convenient controlling switch connected with galvanized sheet iron resistance coils placed in the forward end of the boat. By means of this switch and the circuit breaker, the apparatus is under control and variations in speed are obtained. The pumps as well as screws can be geared to the motor.

The boat carries ordinarily two men, who have remained in her 1 hour and 45 minutes without inconvenience. One of these men acts as pilot ; the other as his assistant and relief. There are two wheels to handle ; one for the rudder, the other for changing the angle of the screws. Mr. F. L. Perry, of the *Western Electrician*, in the course of a most interesting article in that journal,[1] reports spending 35 minutes in the boat, closed up, and chiefly under water ; and mentions a test on May 24, 1892, when Mr. Baker and his assistant were sealed in the boat, partly on the surface and partly below it, 2 hours and 44 minutes. The official "Notes on the Year's Naval Progress," for 1892, issued by the U. S. Navy Department, in speaking of this boat says : "She was frequently submerged, retaining an even keel below the surface, and answering readily to the requirements of the pilot. * * * * It is the opinion of the Bureau that the problem of submarine navigation and attack is approaching a solution and will play an important part in naval defensive warfare of the future."

1. *Western Electrician*, June 4, 1892.

In April, 1892, the submarine electric boat "Audace" was launched by Migliardi Brothers, of Odnito-Vene, at Foce, Italy, to the order of the Roman Company, for fishing and the recovery of treasure, etc., from the bottom of the sea. She measures 8.50 metres long, 3.50 metres in height, and is 2.16 metres maximum beam. She is divided into compartments and is built entirely of steel, having an ovoid form in a transverse direction. She is driven by an electric motor, actuating a screw propeller. It is stated that she can descend more than 300 feet; but the manner in which she does this, as well as the various other features of her operation have been kept a profound secret. The boat will accommodate four or five passengers, and can remain under water with them for six consecutive hours. She has one of her compartments fitted with a door in such wise that through it divers may carry on work in recovering treasure, repairing hulls, fishing for pearl oysters, and other like pursuits.

CHAPTER VII.

DIRIGIBLE ELECTRIC "TORPEDOES" FOR WARFARE AND
LIFE-SAVING.

37. While they cannot be classed strictly as boats, electric dirigible torpedoes are not to be overlooked in a general treatment of the subject of electrical navigation, as they are, practically, electric craft without a crew on board. The advantages of being able to control these torpedoes from the shore or shipboard, are obvious. Some idea of the havoc possible with floating torpedoes may be formed from the fact that toward the close of our Civil War, in 1865, in Mobile Bay, within the brief space of two weeks, no fewer than five Federal gunboats, two of them heavy double-turreted monitors, were totally destroyed by coming in contact with buoyant torpedoes ; and that a large launch was also blown to pieces with its crew. If such effects were possible with uncontrolled torpedoes, it would seem reasonable to infer that infinitely greater injury could be inflicted on the largest ships of an enemy if the torpedoes were controlled or dirigible, and were also automobile. Such types are now extant and successful. The earlier forms were electrical only in the sense of having their steering mechanisms governed electrically. The Lay-Haight torpedo has its rudder actuated by an electrically controlled gas engine, wires running from the operating station to the torpedo through an insulated cable. The Patrick torpedo is also governed in its actions by electricity, by means of a two-wire cable used with 80 Bunsen cells in series. It is our object, however, to deal with the types in which electricity is also the motive power. That which is illustrated in Figs. 40 and 41 is known as the Sims-Edison. The engraving shows the torpedo in section. It consists of a cylindrical hull of copper, with conical ends, and is supplied with a small screw and a rudder. The hull, carrying the dynamite section,

controlling cable, electric motor, and steering gear, is supported at a submerged depth by an indestructible float attached to the hull by an upright steel stanchion. As the boat moves, only the top shell of this float is visible, carrying little rods showing signal flags or balls. This float, against which the Gatling and Hotchkiss guns of the enemy would be vigorously directed, is said to be entirely impenetrable. The hull and float are protected from cables, ropes, or other obstructions by a sharp steel blade set at such an angle as to make the boat dive under or cut away the obstacle.

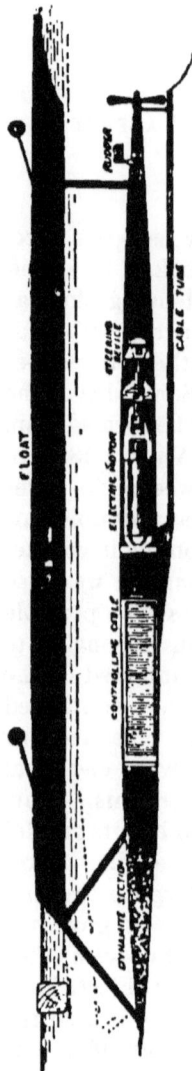

The electric current from the dynamo on shore is conveyed to the torpedo by a cable stored in one of its sections, which is paid out as the torpedo proceeds on its errand. The operator from his station on shore or on shipboard can at will start, stop, or steer the torpedo to port or starboard and explode the charge, which can also be arranged to explode by contact if desired, and he receives notice when the hull or blade meet with any obstruction, together with the magnitude of the same, thus making sure of the proper moment for explosion.

Steering is effected by a powerful electromagnet, into which is switched the main current by means of a polarized relay actuated by the current of the shore battery. Two keys control the relay in the boat, and the rudder is thus thrown from side to side.

The different sections are connected together by gun metal rings with wedge-locking pieces; in other words,

Fig. 40.—The Sims-Edison Electric Dirigible Torpedo.

they screw together with the help of large spanners, and the joints are made water tight with India rubber rings. The act of joining the sections makes electric communication where necessary, by means of spring contacts, so that no jointing of wires is required.

In No. 1 section is seen the charge, occupying the space from the nose as far as a supplementary bulkhead, through which the primer case is screwed into the centre of the charge. At this bulkhead there is a joint, which is, however, seldom broken. The charge carried varies, but may be taken as about 500 pounds of gun cotton or other explosive, which is fired by means of an electric detonator, and not by percussion. No. 2 section is fitted with a lid,

Fig. 41.—Trial of Sims-Edison Torpedo.—Launched from Shore.

which takes right off, so as to allow of putting in the coil of cable, and a tube comes away from it under the torpedo, through which the cable is carried away clear of the propeller. This section is necessarily open to the water. The cable, of which about 7,000 feet is carried, is wound closely on a spindle, layer over layer, and when this is done the spindle is withdrawn, leaving the coil held between two end plates which are held together by four rods. The inner plate is then drawn out, and the operation is repeated in such a manner that after the second winding, the cable on being drawn out from the centre, comes out straight, pliable and free of kinks. It is now placed in its receptacle, the leads in the end are connected through the after bulkhead, and the inner end is rove through the tube and connected when required to the dynamo lead through the switchboard. The cable, which is very flexible, is about ¼ inch diameter over all, and contains two concentric conductors highly insulated. The outer conductor carries the dynamo current for working the motor, and the inner carries current from a secondary battery, for operating relays in connection with the steering gear.

The after part of section 2 is bulkheaded off to form a watertight compartment, which contains a relay for sending the motor current through to the charge. Section 3 contains a two-pole series-wound motor of Edison make, which runs at about 1,600 revolutions per minute, and with a current of 25 amperes at 1,150 volts develops about 33 horse-power. At the after end the shafting is geared down so as to give about 800 revolutions for the propeller. No. 4 section contains the shaft, which turns in "metaline" bearings, so that no oil is required in any part of the torpedo. A simple clutch connects this shaft with that in section 3, but they are insulated from one another by vulcanized fibre. This section also contains the steering mechanism, which consists of two electro-magnets for working the rudder one way or the other, and a relay for sending the main current from the motor to either of these electro-magnets on its road to the frame. On the top is carried the rudder, which, when not drawn over either way, is kept straight by the motion of the torpedo through the water. On the end of the shaft is keyed a gun metal right handed propeller 30 inches in diameter, which for ship use is fitted with a guard to prevent fouling the cable after launching. The sloping stay at the bow is made perfectly sharp so as to cut through obstructions, but failing to do this, the torpedo dives under, and when clear comes again to its former level. The vertical rods on the float are so constructed as to hinge back in such a case, and to spring up again when clear. The total weight of the torpedo ready for service may be taken as 1½ tons, and the floating power is all in the float, which for service is filled with cotton, so that if riddled with shot it would still leave a sufficient margin of buoyancy.

The current for operating the motor is produced by a

continuous current dynamo, shunt wound, capable of producing a current of 32 amperes at 1,200 to 1,300 volts. The current for working the steering relay is taken from a small box of secondary cells, giving current at 50 volts. The torpedo is worked by sending in any given direction the dynamo current, which can be switched on or off as desired, and also reduced or increased, by means of a set of resistances in the shunt. When the charge is to be fired, the main current is reversed by means of a suitable switch. This acts on the relay in the after end of section 2 and the current goes through to the charge. Just before switching over, the current is reduced by means of the resistances. For steering, the current from the secondary battery is sent through in one direction or the other by means of a suitable switch, and so acts on the relay as to send the . main current to either of the two electro-magnets which work the rudder.

The return for both currents is by water through the frame of the torpedo. To avoid a chance of premature explosion, a safety plug is attached to the switch, which must be taken out before it can be reversed, and in later torpedoes there is an arrangement to prevent the circuit being completed through the primer until the motor has made a certain number of revolutions. Near the switch-board at the directing station are placed a voltmeter and ammeter, to show what current is going away to the torpedo, and they also indicate at once when the torpedo has met with any obstruction, owing to the sudden extra work thrown on the motor. When working the torpedo from a ship in motion, the cable is to be paid out from the vessel as well as from the torpedo as required, so that in no case will there be any drag on the cable.

This torpedo has been adopted by the U. S. Army for coast defence. The army reception trial at Willet's Point in 1891, was made with a torpedo 31 feet long, 25 inches in diameter, carrying 2 miles of cable, and having a capacity for an explosive charge of 500 pounds. It was manœuvred at will, attaining a speed of 20 knots an hour or two knots in excess of the contract, although it is said[1] that in more

1. *Annual of the Office of Naval Intelligence*. Gen. Inform. Series, No. XI., July, 1892, p. 128.

recent tests, the same high rate had not been maintained. This shortcoming is attributed to poor insulation of the wires, a defect it should be easy to overcome.

This torpedo has also attracted considerable attention in Europe; and Fig. 42 illustrates some interesting experiments made in England to launch and manipulate it from a vessel under way as well as from a fixed point on shore. A tumbler frame fitted on rollers on an overhead track holds the torpedo, and the detaching is accomplished automatically when the torpedo is clear of the ship's side, it entering the water with an initial impetus which assists it to clear the

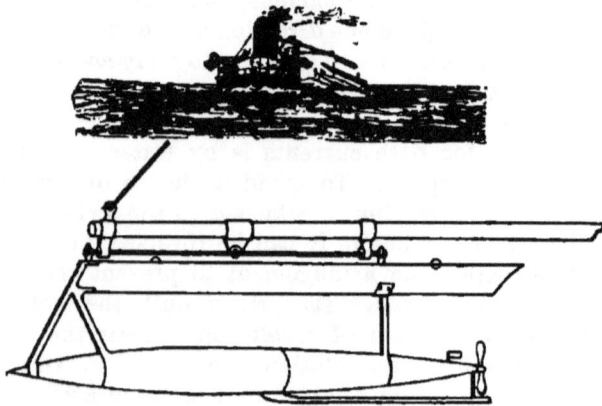

FIG. 42.—TRIAL OF SIMS-EDISON TORPEDO.—LAUNCHED FROM SHIP.

vessel. The final trials took place in Stokes Bay, off Portsmouth, in 1892, when the "Drudge" running at 4 knots an hour launched the torpedo quite successfully. The torpedo was then accurately manœuvred from the ship and ran out 7,000 feet of cable in 4 minutes and 10 seconds, or at the rate of 18.1 knots an hour.

38. Another electric dirigible torpedo is the Nordenfelt, of which a successful trial in England was reported in 1888. It is cigar shaped, moves 8 feet below the surface, and has two floats indicating its position to the manipulator. Its

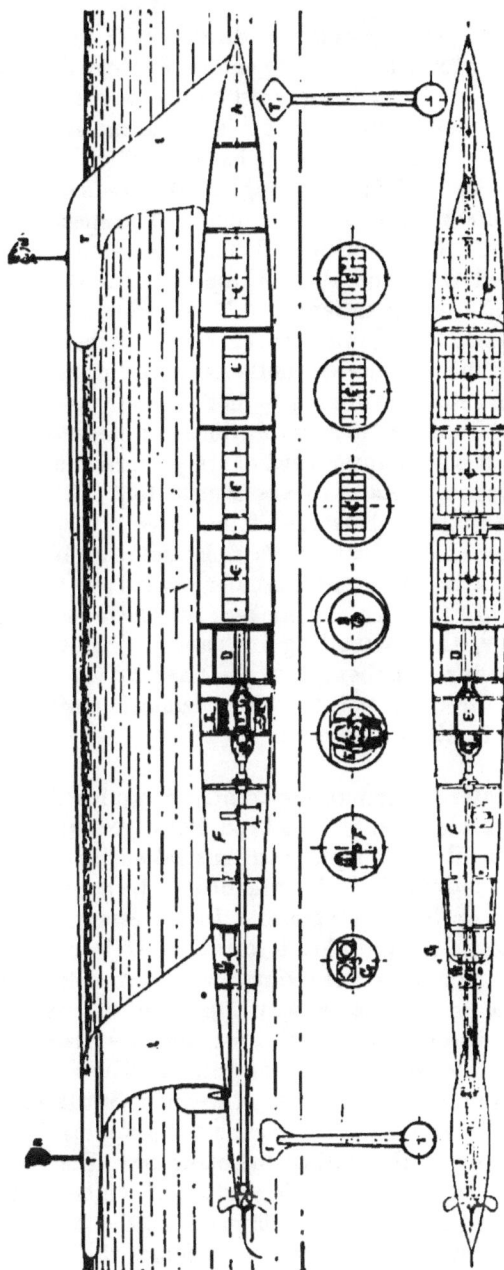

FIG. 43.—THE NORDENFELT DIRIGIBLE ELECTRIC TORPEDO.

length is 35 feet; maximum diameter 29 inches; total weight ready for action, 6,200 pounds; explosive charge 300 to 500 pounds. It carries its own motive power, propelling and steering apparatus and cable. The motive power is furnished by 120 cells of storage battery, which will develop 18 horse-power. The motor weighs 780 pounds and drives a screw at 1,100 revolutions per minute. The speed obtained is 14½ knots. The steering is done by a balanced rudder, manipulated from the shore through a three-core cable, 3,000 to 4,000 yards in length. When tested, the torpedo has run nearly 2 miles in three successive trips. Mr. Nordenfelt had in hand more recently a similar torpedo, to carry 180 storage cells, developing 34 horse-power, and giving a speed of 16 knots at 1,500 revolutions per minute; but the writer has not seen any record of its performance.

Fig. 43 shows details of the Nordenfelt, while Fig. 44 contrasts the general design of the three types, the Sims-Edison, the Lay-Haight or Patrick and the Nordenfelt. In Fig. 43, A is the charge chamber; o, c, c, o, the storage batteries; D, the cable chamber; E, the electric motor chamber; F, the controlling instrument chamber; G, the steering power chamber; T, T, the fins; $t\ t$ the fin plates; N, N, the directing points, which contain electric lights for service in night operations.

39. An interesting but apparently complicated torpedo of the dirigible type is that brought to general notice in 1890 by Mr. Read Murphy, an Australian, and called by him the "Victoria." It is intended to be operated either from shore or from shipboard. In all essential respects the weapon is a Whitehead torpedo. It is of the same general shape and construction, and is propelled by compressed air in the same way. But to the Whitehead equipment there are added some new features of great importance by which it is controlled. The steering devices are found in both types of weapon, but in the shore torpedo there are provided means for stopping, starting, and exploding at the will of the operator. The shore torpedo is the larger and more important of the two. It is 24 feet long and 21 inches in diameter at the largest part. The head carries the charge; next comes the compressed

air chamber; behind this is a chamber in which are coiled
1,200 yards of electric cable inclosing the insulated copper
strands; further aft again is a chamber in which are three
electrically controlled spring motors, one for working the
vertical rudder, the second for the air valve which controls
the propelling engine and allows the cable to pay out from
the torpedo when at full speed, and the third for exploding the torpedo and also for bringing it to the surface. It
is through the agency of the cable and the motors that the

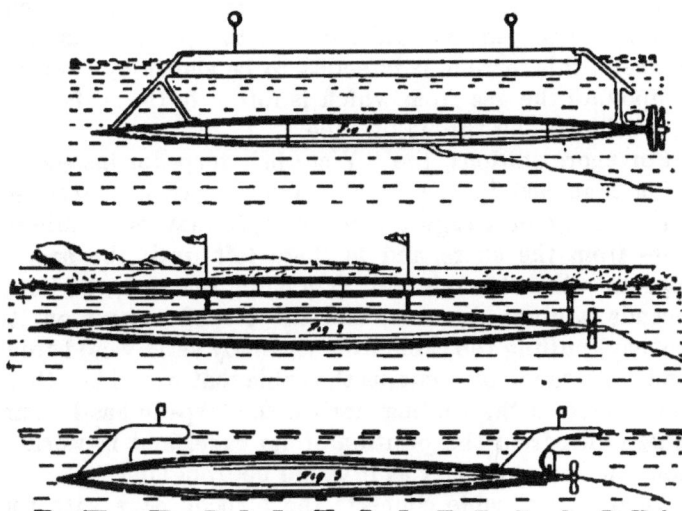

FIG. 44.—THE SIMS-EDISON, LAY-HAIGHT AND NORDENFELT
TYPES OF DIRIGIBLE TORPEDOES.

torpedo is controlled. The 1,200 yards coiled within it,
however, do not represent its range. An additional quantity is coiled at the place where the operator is situated,
and it is intended that this shall be drawn upon first. As
the speed of the weapon is under control, it is seldom
advisable to launch it at full velocity. The amount of
power required to drive it at full speed for, say, half a
mile, would propel it at several miles at half speed, and
hence it is usually wiser to send it away at a moderate rate
until within easy shooting distance, when full speed can

be given if desired. During the first part of the time the
torpedo would be dragging its cable behind it, not a very
serious matter, since it only weighs 41.110 grammes per
yard in the air, and when greased it will leave the torpedo
22.312 in water. But when the air valve is opened wide to
give full power, the coiled cable is released and pays out
from the body of the torpedo, thus obviating the drag, and
giving 1,200 yards free run. There is also a device provided whereby, should the shore end of the cable become
fouled, or offer too much resistance to the motion of the
torpedo, the clip which holds it is tripped, and the cable
within the torpedo is paid out. In the present case the
course of this torpedo is intended to be shown by Holmes'
compound, the gas from which is forced to the surface by
the rush of water through a tube.

For shore stations, the "Victoria" torpedo, instead of
being launched in the usual manner, may be deposited
with a buoy in a cage under water, it may be a mile or
more from the shore, and is there left until the enemy
appears, when it is released. On being released, the buoy
ascends a given distance, and the torpedo starts on its
journey, pulling the cable from the buoy as it would from
land. As the buoy contains the cable that would be otherwise wound at the sending station, the torpedo has its run
of 2½ miles from the position of its cage, and is worked
from that point exactly as it would be from land.

An ordinary cable connects the torpedo cage with the
operating station on shore. This cable contains the three
strands already described, of an extra cross-section to carry
the extra current of electricity ; and two extra strands, one
of which enables the torpedo and its buoy to be released at
pleasure, and when not required for this purpose is connected with an ordinary electric bell. The second strand is
connected with electric cells, so that if the torpedo or the
cable is interfered with by the enemy, or hurt by mis-
chance, the water will connect the two strands so that the
bell will ring, and the officer instantly be apprised. If,
however, all goes right, he can, by a touch of the key, open
the cage and liberate the torpedo, which will rise as described ; its engines can then be set in motion, and it can be
controlled and steered at will. The feature of interest,

electrically speaking, is that each spring motor in these torpedoes is connected with a small electric motor, by means of pulleys and clutches. No record of any use of the "Victoria" dirigible torpedoes is at hand. Their mechanism seems to be most unnecessarily complicated.

40. We have thus far been considering dirigible craft of the torpedo character simply as weapons of naval warfare, but it is obvious that they constitute an admirable means of saving life at sea either as a projectile with life lines to aim at a wrecked or stranded ship or as an apparatus that can be directed towards an overturned boat or to persons struggling in the water and needing a buoy. Indeed an electric life boat of this description has been devised in England by Mr. J. Hibberd, whose plan is illustrated in

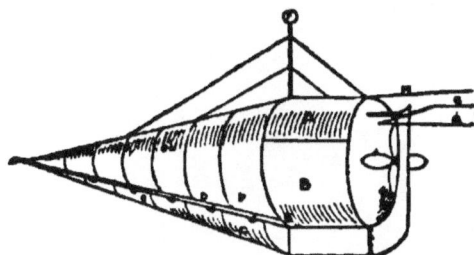

FIG. 45.—HIBBERD LIFE-SAVING DIRIGIBLE ELECTRIC TORPEDO FLOAT.

Fig. 45. A, A are the air chambers ; B, the electric motor ; C, sand or a grapnel ; D, rod to open the bottom to release sand, etc.; F, an electric light ; G, G, guide lines, and H, communication line. This float can be sent from shore to ship or vice versa. The object of loading one of the spaces with sand is to keep the float under water till it reaches the object aimed at, when, the sand being dropped, the float rises, and can be secured.

An electrically lighted life buoy has been invented lately by Capt. Melter, and some trials were lately made with it at Kiel on board the German war vessel "Worth." The buoy was thrown overboard when the vessel was proceeding at a speed of about 16 knots, and for about 12 seconds it was lost in the eddy current caused by the twin screws of the vessel, but then reappeared.   It is stated that the experiments resulted so successfully that it is probable the new life buoy will be adopted generally in the German navy, and there seems no doubt it will be found of great value at night time.   It is evident that if such a buoy were dirigible, as it might easily be, its value would be incalculably increased, not only for use on ships but for utilization by shore patrols.

CHAPTER VIII.

SOME GENERAL CONSIDERATIONS ON ELECTRIC LAUNCH REQUIREMENTS.

41. In the course of the preceding chapters a great many features in the construction, operation and maintenance of electric launches and boats have been touched upon; note has been made of many of their advantages and conveniences, and some suggestions have been offered of means for improving and developing such work as the service involves. We now supplement such information by two tables furnished by Mr. J. C. Chamberlain, E. E., giving data, in harmonious relation and of an interesting nature as to size, capacity, speed, cost, etc. The first category includes yacht tenders, for yachts that have a charging plant, and pleasure boats to be charged for regular stations:—

Length over all—ft. and in.	21	0	23	0	25	0	24	0	27	0	30	0	35	0	40	0
Beam " "	5	2	5	4½	5	6	5	2	5	4½	5	8	6	2½	6	9
Freeboard " "	1	1	1	2½	1	4	1	1	1	2½	1	8	1	5½	1	6½
Draught " "	1	10	1	11½	1	11½	1	11½	2	1	2	2	2	2	2	4½
Speed-Miles per hour for continuous run of 3-4 hrs.	5½		5½		6		5½		6		6½		7		7¾	
6-7 "	4¾		5		5½		5		5½		6		6½		7	
8-10 "	4½		4¾		5¼		4¾		5		5½		6		6½	
Speed rate for short spurt .	7		7½		8		7		8		9		10½		11-12	
No. of batteries stand. type	20		28		40		24		36		52		68		100	
Charging current—volts .	50		70		100		60		90		130		85		125	
amperes	25		25		25		25		25		25		50		50	
No. hours to recharge .	4-6		4-6		4-6		4-6		4-6		4-6		4-6		4-6	
Cost per hour to recharge batteries—cents . .	7½		10½		15		9		13½		19½		25½		37½	
Seating capacity . .	10-12		12-15		15-18		12-15		15-20		20-25		25-30		30-40	
Prices—Stand. type, planked with selected cedar, decks of mahogany, entire hull and finishings handsomely polished and varnished .	$1600		$1800		$2150		$1650		$1875		$2250		$2700		$3650	
Same type of boat, painted hull, pine decks, calked, handsomely finished .	1500		1675		2000		1550		1700		2100		2550		3350	

The table is, it is thought, the more valuable for the prices it contains, as it affords an idea of the utmost expenditure involved in buying and operating electric boats. The next table, from the same authority, deals with regular electric launches, running from 18 up to 70 feet in length.

ELECTRIC EQUIPMENTS FOR LAUNCHES.

CONSISTING OF MARINE TYPE OF STORAGE BATTERIES AND MOTOR, ALSO SPEED CONTROLLER AND REVERSING SWITCH, CONNECTING WIRES, BINDERS, INSULATING TAPE AND PROPELLER WHEEL.

No. of equipment . .	1	2	3	4	5	6	7	8	9	10	11	12
For boats of L. W. L.—feet .	18-21	22-25	26-29	30-33	35-39	35-39	40-44	40-44	45-49	50-55	50-55	60-70
Speed—Miles per hr. for continuous run of 3-4 hrs.	5¼	5¾	6½	7	7¾	8	8¼	8½	9	9¾	10	11½
6-7 "	4¾	5¼	6	6½	7	7¼	7½	7¾	8¼	9	9¼	10½
8-10 "	4½	5	5½	6	6½	6¾	7	7¼	7½	8¼	8½	9¾
Speed rate for short spurt .	6½-7	7-8	8-9	10-11	11-12	11-12	12-13	12-13	13-14	14-15	14-15	15-16
No. batteries in each equipm'nt	20	36	52	68	100	104	132	136	168	200	208	264
Type of motor and controller	No. 2	No. 4	No. 6	No. 8	No. 12	Two No. 6	No. 16	Two No. 8	No. 20	No. 24	Two No. 12	Two No. 16
Charging current—volts .	50	90	65	85	125	130	85	90	105	125	130	85
amperes .	25	25	50	50	50	50	100	100	100	100	100	200
No. hours for full recharge	4-6	4-6	4-6	4-6	4-6	4-6	4-6	4-6	4-6	4-6	4-6	4-6
Cost per hr. to recharge—cents	7½	18	19½	25	37½	39	51	54	63	75	78	$1.02
Total weight—pounds .	1050	1740	2570	3240	4625	5170	5955	6540	7590	8940	9710	11850
Price, boxed F. O. B. N. Y. City	$850	$1170	$1500	$1830	$2650	$3010	$3310	$3680	$4210	$4850	$5470	$6390

Note will be taken of the fact that the greatest length here provided for is 70 feet; and the greatest number of cells, 264. These are figures far beyond the ordinary to-day, in electric launch work, but they were actually exceeded in the remodeled and enlarged "Electron," rebuilt on the Hudson, in 1890, by Mr. James Bigler, from 36¼ feet to 76 feet. Her original equipment of batteries was 200 cells, but in the rebuilt boat this number was increased to no fewer than 376. This boat had decks and a cabin, and was operated at Atlantic City, carrying passengers for fare from shore to sea, two or three miles, and back. But her use was attended with patent and other litigation, pending the settlement of which she has been more lately operated as a steamboat. Her return to the ranks of electrical boats is awaited with interest, and it is to be regretted that in this case as in so many others of late years, where storage batteries have been concerned, fruitless legal squabbles have been allowed to interfere with the legitimate advance of the art.

42. It will interest many readers to know what are the elements entering into the design and construction of such boats, and a few points dealing with such topics are now given, with the assistance of Mr. F. Reckenzaun, who has so long devoted his attention to this subject.[1] The first item to be taken into account is the hull, which, however, need not materially differ from that of a steam or naphtha launch. The shipbuilder's work remains the same, only the interior or joiner's work requiring adaptation to the different nature of the equipment. By this is not meant, however, that any sort of hull will give satisfaction. Before we have it built it is well to know what we are going to put into it—the kind and size of motor and the number and size of battery cells. Space is limited on all sides, and the features of the hull, motor and battery should be carefully considered in their mutual relations.

Let us assume that we wish to design our own boat. After we have roughly modeled the lines, either in accordance with the shipbuilder's practice, or to suit a special fancy, it will be of advantage to lay down upon paper the cross sections, taken at suitable distances apart from bow to stern, a longitudinal section and a plan. We have but a single straight line in an ordinary launch hull, and that is the keel. The rest are all curves of various character. These curves and the dimensions of the motive power outfit

---

1. The accompanying sections are largely based upon an admirable contribution of Mr. Reckenzaun to *The Electrical Engineer*, New York, of Aug. 13, 1890.

should mutually agree, or else they are likely to prove awkward when the apparatus is being put in place, and a sacrifice of some kind would be the probable consequence, which, perhaps, a very slight deviation from the lines of the model might have avoided. The hull should be substantial throughout to withstand the strain from the weight of battery and motor when in rough water. The joiner's work will include a battery receptacle or receptacles of the required dimensions. A trough placed directly over and along the keelson, with the seats arranged on top of it, has the advantage of giving the boat maximum stability, since the centre of gravity will then fall near the keel, below the water line. The passengers sitting back to back in two rows along the centre, will add to this advantage, while at the same time they may have an unobstructed view in front of them. Another method consists in placing the battery in a similar trough, laying the floor over it, and arranging the seats above this floor along the sides or across the hull. Access to the battery can be obtained in the first instance by removing the top of the seats, and in the latter through trap doors in the floor. Again, another way of distributing the cells consists in arranging the battery receptacles along the sides of the hull, with seats on top of them. A combination of the above methods may be effected to suit preferences. It is well to remember, however, that one of the advantages of the electric launch is that its stability may be made to exceed that of any other launch by a judicious distribution of the weight of the propelling apparatus.

As to the material of the hull, wood is preferable to steel or iron when it is considered that acid is to be carried on board, although by a suitable construction of the cells and receptacles leakage or spilling can be prevented under ordinary conditions. These remarks apply, however, to ordinary boats. It has already been pointed out that submarine boats should have steel hulls, while some have been built of phosphor bronze.

43. The next element to consider is the motor, which must embody high efficiency, with special compactness, low speed and reasonably light weight.

It is usually desirable to put as large a power outfit into.

a boat as can be conveniently placed there and looked after, and since the battery requires the largest amount of space, the most suitable place for the motor is in the stern, as far back as possible. The shape of this space then will generally determine the selection or construction of the motor. The cross section in an ordinary launch hull resembles the shape of the lower part of a heart, diminishing in width toward the sternpost. The available base area for the motor being triangular, the base must be narrow. With an armature of about double the length of its diameter and the field magnets crowded around it to suit the lines of the hull, we have a motor that can be placed without unnecessarily encroaching upon space desirable for batteries or passengers. A low armature speed (say 500 to 800 revolutions per minute) will admit of coupling the motor shaft directly on to the screw shaft without necessitating excessive fineness of pitch in the screw. We have here an ideal method of transmission. There is no lateral strain on the motor bearings, while the thrust bearing, interposed between the motor and the screw, takes up the longitudinal strain.

The conditions of load in a launch motor are analogous in the main to those of a fan motor, but more particularly to those that would be encountered in an air ship or flying machine, the following characteristics being observed : The load consists in the resistance offered by the water to the motion of the screw. The movement of the latter is independent of its support—the boat in the present case—and its effect will be either to set in motion the medium (water) in which it moves, if the support is fixed, or to propel the latter if it is free, in consequence of the inertia of the medium. Hence the force required to start a boat is merely that required to overcome the inertia of the body of water affected by the screw proper and is independent of the inertia of the boat.[1] The motion, being first imparted to the water, is gradually transferred from the latter to the boat, until a point of equilibrium is reached, determined by the resistance encountered. It will be seen from this that even if we throw the full load upon the motor at once (as is usually done), the difference between the starting effort and that

---

1. The subject of screws is treated in a later chapter at more length.

required to maintain the final speed of the boat is so small as to be entirely negligible in the determination of the capacity of the motor. Speed once being reached, the load will remain constant on a straight run, while it will slightly increase on turning the vessel about, until a straight course is resumed. In running against a current, the load of the motor is smaller than in still water, while it is greater on running with the current. Under the conditions ordinarily met with, this difference is, however, but slight. Automatic governing devices are obviously not required, unless we were to consider an equipment for a large sea-going vessel. If it is desired to get two or more different rates of speed, the battery may be split up into a corresponding number of sections, by means of a special switch connected in series or parallel. The brushes should admit of reversing the direction of the armature movement, but their "lead" is best adjusted for forward motion of the boat, unless a double set of brushes is employed, with reversing lever to engage one or the other as required. A suitable switch inserted between motor and battery serves for starting, stopping and reversing the motor.

44. The storage battery, on account of its superior fitness, is universally employed in connection with electric launches at present. Without entering here upon details of construction, we will consider the features to be dealt with in its application. The first question confronting us is that of bulk and weight. The manner of disposing of the battery has already been touched upon in considering the hull. Being composed of a number of small units, there need be no difficulty in distributing it. An ordinary launch hull can well carry all the *weight* corresponding to the *bulk* of battery which can conveniently be placed with due regard to accessibility. We will, for convenience, take into consideration a battery of a well-known type,[1] designed for portable and locomotive purposes, occupying, per cell, 0.23 (solid) cubic foot of space (box, plates and all) and having a capacity of 150 ampere-hours, or about 290 watt-hours, at a

---

1. The cell here dealt with has already undergone considerable improvement, but the figure will serve as a conservative estimate. The subject of batteries is resumed in a later chapter.

discharge rate (normal) of 25 amperes or an average of about 48.32 watts; its weight being about 40 pounds. Reduced to unit cubic foot, we have: Weight, about 175 pounds; capacity, about 1,260 watt-hours, with normal discharge rate of 210.1 watts per cubic foot. The displacement per cubic foot of battery will then be equivalent to ($\frac{175}{62.5}$ =) 2.8 cubic feet of (pure) water; at this rate, allowance, for weight of battery, must be made in determining the water line. With ordinary launch hulls, the average battery load that can be carried conveniently represents about one-third of the total actual displacement in tons, including passenger load. Any smaller proportion may, of course, be applied, with correspondingly reduced results in capacity. The capacity of the battery and motor are considered mutually; for maximum effect the former guides the calculation. With the type of battery above assumed, if it is to be worked at "normal" rate, the capacity in electrical horse-power of the motor required will be equal to the number of cells multiplied by 0.0047, the working rate in electric horse-power per cell, or to the number of cubic feet of battery multiplied by 0.2816, the corresponding constant per cubic foot. For a rough preliminary calculation, on the basis that the weight of battery represents one-third of the total weight, we have,

$$\text{Capacity} = \frac{D}{3} \times 3.604 \text{ electrical horse-power;}$$

where D = total displacement (weight) in tons and 3.604 the working rate of battery in electrical horse-power per ton (12.8 cubic feet). The duration, T, of the run in hours for one charge of battery will be:

$$T = \frac{C}{R} \text{ hours, } C \text{ denoting capacity of battery in watt}$$

hours and R denoting rate of delivery in watts.

Since the power required to propel a vessel varies as the cube of the speed, and since the duration of the run varies inversely as the power (rate of delivery), it follows that the *mileage* covered by one charge of battery will vary inversely as the *square* of the *speed*. In practice, due allowance is to be made for the characteristics of the motor and for a falling off in the total output of the battery when pushed to a high rate of delivery. Where a maximum of speed is to be effected, the battery should have a maximum of active surface and a minimum of internal resistance, to facilitate a heavy discharge without an excessive drop of potential. Special care should be taken to render the cells acid tight, by the use of suitable covers, etc. Spilling may also be avoided by preparing the electrolyte in a suitable manner. The jelly electrolyte invented by Dr. P. Schoop offers in this respect a remarkable advantage.[1] It is also advisable to line the battery receptacle with some acid-proof material, preferably an insulator, and to provide a bed for the cells to stand on containing a substance capable of absorbing and neutralizing acid. All wires or cables employed about the boat should have a good acid and salt-waterproof insulation.

The question of charging facilities has already been pretty fully discussed, but cannot be too exhaustively considered by the owner of an electric launch. It is believed that the introduction of dynamotors (or continuous current transformers), by means of which any direct current, as from railway or lighting circuits can be raised or lowered to the potential required by the battery, is destined to prove of incalculable service in promoting the electric launch industry. But let us assume, in order to present the necessary calculations, that we have a 40-cell launch, cells being of 150 ampere hours capacity, and that 110 volt current is available from some everyday incandescent lighting plant. The difference of potential required at battery terminals (cells in series) would then be 40x2.5 = 100 volts at the finish—if a constant current, 25 amperes in this case, is to be maintained. To reduce the initial electromotive force of the circuit, we must then introduce a resistance of

$$R = \frac{110-80}{25} = 1.2 \text{ ohm}$$

at the start and gradually reduce the same as charging goes on to the final minimum of (approximately)

$$R = \frac{110-100}{25} = 0.4 \text{ ohm.}$$

1. See later chapter on Storage Batteries, etc.

Or, we may apply a constant electromotive force of about 90 volts (2.25 volts per cell) by inserting a constant resistance of

$$R = \frac{110-90}{25} = 0.8 \text{ ohm,}$$

in which case we will receive a heavier current at the start, reducing itself (in consequence of the increasing counter electromotive force of the battery) gradually to a minimum at the end, the average being the same as in the other case. The latter method may be preferable; the current will decrease in proportion to the facility with which the gases can be absorbed by the plates, while the results in time and efficiency remain practically the same, and constant attention is rendered unnecessary. If a boat is to be charged, the battery of which, connected in series, requires a higher electromotive force than that available, we merely need to split it up into two or more equal series to get within the required limit and then charge these in parallel with a proportional current, adjusted as above.

The battery may, of course, be charged either on the boat or may be removed for that purpose. While the former method is ordinarily practiced, it is obvious that in order to avoid delay, a freshly charged battery may be substituted for the exhausted one. With suitable facilities for handling the batteries, such as a hoisting crane, or equivalent device for lifting and lowering the cells into and out of the boat, tables to receive the cells for charging, suitable cell crates with connections and lifting attachments, etc., the work of exchanging the batteries could be effected promptly and efficiently for a whole fleet engaged in continuous traffic. It is evident, of course, that a dynamotor would dispense with all such handling of the batteries.

CHAPTER IX.

CANAL BOAT PROPULSION : HISTORICAL.—ERIE CANAL.

45. The invention of Canals or artificial waterways can be carried back to the earliest ages of civilization. Various historians credit the ancient Egyptians with having first devised and constructed artificial interior waterways as early as 1500 B. C. The Chinese were also among the first to build canals, and the great Imperial Canal of China about 1,000 miles in length is still in existence. The canals constructed · by the ancients did not, however, resemble our modern canals ; they were in fact simply large ditches dug through level stretches of country. When it was necessary to pass from one level to another, the boats, which were comparatively small, were raised or lowered by means of inclined planes. The Chinese were perhaps the first to make use of such devices.

Although various canals were built throughout Europe by the Romans, canal building did not make any great progress until the invention of the canal lock as used at the present day. With this improvement the canal became a most simple, effective and economical source of interior transportation. In fact, until the advent of the railroad, the canal was the great carrier of all classes of products and goods through level countries where no natural waterways existed. A very large number of canals have therefore been constructed and remain in use throughout the world at present.

Exact statistics as to the cost, mileage and use made, of the canals of the world are not obtainable. The latest authentic figures we can compile give Europe between 12,000 and 15,000 miles of canal, including England with 4,700 miles ; France, 3,000 ; Germany, 1,250 ; Holland, 930 ; Belgium, 540. In the United States, according to the report made by Special Agent T. C. Purdy for the Census of 1880, there are 4,468 miles of canal that have cost $214,000,000. Of these, 1,953 miles were abandoned, in 1880, and of the remaining 2,515 miles, quite a large proportion was not paying expenses.

46. Navigable canals may be divided into two classes: Ship Canals, which can be navigated by seagoing vessels of large draught and tonnage; and Barge Canals, which are generally rather shallow and narrow and only permit the use of barges or lighters thereon. The first class is confined to a few examples, such as the Suez, Manchester, Welland, Corinth and various ship canals in Holland. They have been of immense value to commerce and will grow in number.

It is with the second class, or *barge canals*, that we shall particularly deal. Of this class there are very many in existence, but to a certain extent the barge canal has unfortunately fallen into disuse in those countries where the railroad has become a serious competitor. Few have been built within the last fifty years, while thousands and thousands of miles of railroad have been constructed. Although such canals have to a great extent been superseded by the railroad as carriers of freight, it is acknowledged that they remain one of the cheapest and simplest methods of transportation for large and bulky products.

It is true that the railroad is many times as rapid as the canal, but it must be remembered that a single canal boat of from 100 to 250 tons, such as used on most modern canals, will carry as much material as can be transported by from 10 to 20 freight cars; and the cost of operating a railroad train, carrying a load equal to that carried by a canal boat or number of boats, is so very many times in excess of the cost of transporting the same load by canal boats, that even considering the time element, canal transportation is cheaper for very heavy and bulky loads. To enable the canal to be better able to compete with the railroad, however, it is primarily necessary to improve the present methods of propulsion.

47. Upon the early canals, the boats were pulled by slaves; gradually the horse or mule was substituted for

man power and with the advent of steam various forms of this source of energy were adapted to canal boat propulsion. The majority of the present barges on canals are still hauled by horses or mules.

The rapid advance of electricity and the many things that have been accomplished with this form of energy, have caused it to be suggested at various times as a source of power for canal boat propulsion. It is particularly within the last two or three years that the electrical propulsion of canal boats has become a leading topic of discussion. In this country various state legislatures are about to pass or have already passed bills looking to the improvement of the state canals by the adoption of some form of electrical motive power. Private corporations owning canals have also taken steps to investigate this substitute for the mule, and experiments have been tried, both in this country and abroad, with one or two forms of electrical canal boat propulsion. During the latter part of 1893 experiments were made upon the Erie canal near Rochester with an electrical canal boat. Experiments have also been made in France recently. Of these trials more will be said in later chapters.

48. The largest, most important and most valuable of our American canals, and one which will serve as a good example, is the Erie, connecting the Great Lakes with the Hudson. It has materially helped to build up the internal commerce of our country. This canal was started in 1816 and finished in 1825. Boats were first operated upon it in 1826. It practically cuts the State of New York in two, extending from Troy to Buffalo. As originally built, this canal was about 363 miles long with the following dimensions: Width of bottom 28 feet, width at top 40 feet, depth 4 feet. Eighty-four locks were used along the route, each 90 feet long and 15 feet wide. The boats used were 78 feet 8 inches long, 14 feet 5 inches beam and 3 feet 5 inches draught when loaded to 80 tons. Each boat was towed by one mule.

As, however, the value of the canal became apparent and it served as the great outlet for the products of the west, these dimensions were found inadequate, and it was gradually enlarged to the present dimensions : Width of top 70 feet, bottom 52½ feet, depth 7 feet, and the length shortened to 352 miles.

The locks and structures were enlarged and the boats gradually increased to the present size : 98 feet long, 17½

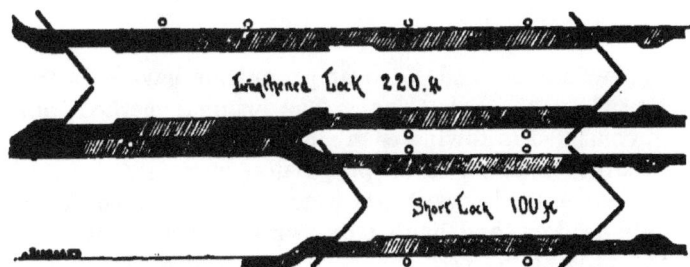

Fig. 46.—Short and Long Lock on Erie Canal.

feet beam and 6 feet draught when loaded to 240 tons. There are at present 71 locks having an average lift of 8 feet and most of them having a length of nearly 220 feet so as to permit two boats coupled tandem to pass through at one time. Some of the locks have not been lengthened and are only 110 feet long, but these will all be changed to the greater length.

Several large aqueducts and embankments on which the canal is carried over rivers, etc., are located along the route. It may also be well to state that there are about 207 bridges

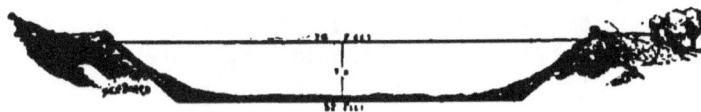

Fig. 47.—Section of Erie Canal.

of various kinds crossing the present canal and that the towing path, which is from 10 to 15 feet wide, crosses from one side to the other very frequently. The illustrations, reproduced from *Seaboard*, slightly reduced, show a standard section of the canal and a general plan of the present locks. (Figs. 46 and 47.)

The Erie Canal cost originally about $7,500,000, but with the improvements this has been increased to about $51,500,000. The canal is only in operation about two-thirds of the year; generally opening from the 1st to the 5th of May and closing between November 30th and January 3d. The average navigable season lasts about 215 days.

49. Various methods of boat propulsion have been tried and used on the Erie Canal. The original method, however, consisted in towing each single boat by a mule on the tow path. This mode of propulsion was used for some time. Afterwards two boats were coupled tandem (called double-headers) and hauled by two or three mules. This is the present method and with the existing large boats

FIG. 48.—ORDINARY MULE HAULING ON ERIE CANAL.

three mules are used (Fig. 48) and an average speed of 1½ miles per hour is obtained through water. Over 95 per cent. of the boats are moved in this way at present. They take about 10 days to go the entire length, including lock-ages.

The necessity for an improved and more rapid method has caused various forms of mechanical power to be tried, and in 1871, the Legislature of New York passed an act offering a prize for an application of steam to canal boat propulsion on the Erie. This brought forth various types of steam propeller canal boats some of which have continued in use to the present day. In 1873 a system of Belgian cable towing was also installed and tried on a level between Buffalo and Lockport, but it was discontinued as it proved unsuccessful. Various other plans were also tried.

There are in use on the Erie Canal about 60 steam propeller canal boats. These steamers are about the size of an ordinary barge, but are of somewhat finer lines, and when loaded to 6 feet draught carry about 180 tons in addition to machinery. They are generally equipped with boiler and simple condensing upright engine, revolving a screw at the stern 6 feet 6 inches in diameter at about 100 revolutions per minute.

The usual modus operandi of propulsion by this method is for the steamer to push one barge, to which it is rigidly coupled, and tow two or more (generally two) boats by a tow line. The boats are all coupled in pairs or double headers. Such fleets or trains attain a speed of about 2½

FIG. 49.—STEAM CANAL BOAT AND CONSORT.

miles per hour and make the trip in a little over seven days. (Fig. 49.)

The trip of a canaler is, however, not finished when the Hudson is reached, but the freight must be transported to New York. The horse or mule boats are made up into large tows and are taken down the Hudson by tugs. The steam canaler attended by its consorts, proceeds down the river under its own power.

There are, therefore, two methods of propulsion on the Erie Canal at present, mule towing and steam propeller. The latter has proved the cheaper method and although the mule boats are in the great majority, it may be said that they do not pay their owners any very great profits. Unless the season's traffic is unusually good, the canaler does not reap a large income in spite of some very hard work.

50. In 1893 the New York State Legislature passed an act authorizing experiments to be made on the Erie Canal with a view of securing a method of electrical propulsion of canal boats that would be superior to the present methods; and, as has been stated, one method has already been tried. This mode of applying electric power consisted in substituting an electric motor for the steam engine used in the present steamer and conveying current to the motor by suitable wires fed from a central station.

Electrical canal boat propulsion is, however, not confined to this particular method. There are various other ways of applying electrical energy to the propulsion of boats in narrow and shallow waterways. Although the methods hereafter described are particularly considered with a view to their application to the Erie Canal, their employment is not in any way limited to any particular canal.

CHAPTER X.

Conditions Entering Into Canal Boat Propulsion.

51. Before taking up the various systems of electrical canal boat propulsion, it will, perhaps, be well to consider what features and advantages any new system of propulsion should possess, and to discuss, generally, the application of such an improved method of boat propulsion on the Erie and similar canals.

The prime object is to cheapen transportation. This can be done in two ways: (1) by increasing the speed of propulsion, or (2) by decreasing the cost of propelling power. The first depends on the canal and the method of propulsion; the latter, upon the nature or form of the propelling energy used. In regard to electricity, it has been claimed that both results could be accomplished by its adoption. Although the adoption of electrical motive power in some form or other will perhaps be more economical than the present methods, increased speed cannot be obtained upon the Erie Canal with any system without changing the present dimensions. With the present depth of barely 7 feet and the boats loaded to 6 feet draught, it happens very frequently that the boats, when going at 2½ to 3 miles per hour, come in contact with the bottom of the canal, which sometimes causes great injury to the boats.

Whether electricity or any other method of propulsion is adopted on the Erie Canal, it is absolutely necessary to deepen the canal before any increased speed can be obtained.

Various plans have been suggested for deepening, the present channel of the Erie Canal. Among the most noteworthy, is a plan advocated by Ex.-State Engineer Sweet for raising the banks and structures of the channel one foot and thereby securing a depth of water of 8 feet. Another plan is advocated by Ex.-State Engineer Horatio Seymour, to bring the depth of water to 9 feet by excavating one foot and raising the banks and structures one foot. The excavation of the bottom would necessitate the replacing of the impervious material composing the present floor. The simplest and perhaps cheapest plan to gain additional depth of water, would, therefore, be to raise the banks as much as possible. It may be noted that a bill has passed the last New York Legislature which authorizes the present depth to be increased to from 8 to 9 feet throughout.

52. There is, however, another consideration. It has been found most economical to operate the boats on the Erie in pairs. This makes the steering of the boats quite difficult, and it will be found impracticable to steer at a very greatly increased speed; but the present arrangement could no doubt be continued at a speed of perhaps from 3 to 4 miles per hour.

The increase of the depth of water would either permit an increase in the present speed with the present boats and methods, or would necessitate less power to propel the present boats at the present speed. Either plan would certainly lower the cost of transportation. The total propelling power necessary for a certain speed would, however, be greatly dependent also upon the mode of applying the propelling power and the cost of furnishing a horse-power-hour of energy at the boat.

All practical electrical systems contemplate the generation of current in central stations along the route of the canal and a distribution of the power over suitable wires to the motor devices. This will necessitate quite a number of transformations from the prime mover to the motor shaft.

That Niagara will some day furnish electrical energy as far as the Hudson is not to be doubted, and the subject is treated fully in a later chapter. It has also been suggested to use water power and the waste weirs along the route of the canal. The want of sufficient water power adjacent to the canal makes this inadequate to the requirements, although no doubt it can be utilized to some extent.

Contemplating merely the various transformations necessary, from the burning of coal, in the boiler, to the delivery of power available for propulsion on canal boats, it seems hardly probable that a horse-power-hour of energy can be furnished cheaper than, if as cheap as, the same amount of power as at present generated by the use of boiler and engine directly upon the boat where a consumption of from 3 to 6 pounds of coal gives a horse-power-hour at the engine.

53. There are, however, numerous other considerations in regard to the application of various forms of electrical propulsion to canals.   The various features that a practical system should possess are :

First.—The system of propulsion should not in any way injure or affect the structures or banks of the present canal in any way.

Second.—It should not necessitate equipping the canal boat with expensive machinery which can only be used where there are corresponding systems of electric transmission available, and which always takes up valuable room.

Third.—The chance of being stalled by breaking down of machinery or lack of current supply must be reduced to a minimum.

Fourth.—Any new method must permit of a continuance of the present methods of propulsion.

Fifth.—If any structure along the canal is necessary, it should be strong and not easily injured.

Sixth.—The labor should, if possible, be less than that at present required.

54. All electric systems of whatever nature, receiving their current supply from a central source of generation or distribution, can only utilize their propelling machinery where there is a source of supply of current.   It has, therefore, been stated that the application of electrical power to canal boat propulsion on the Erie would not be complete as it would not allow the boat to continue down the Hudson by the same power.

This is certainly true for the present, and would be an objection to any system contemplating the equipping of the freight-carrying boat with electrical machinery receiving current from a central generating station.   By the use of storage batteries on the boat, furnishing current to an electric motor revolving a screw propeller, this would not be the case, as the boat would then be self-contained. The application of the storage battery to canal boat propulsion is, however, not yet in a practical shape. The self-contained storage battery boat is a possibility, but will not here be considered at any length.

If, however, some application of the electric motor to the propulsion of these mule barges will lower the cost of propelling them below that of any other means of propulsion, the application of such cheaper method would unquestionably effect great economy in their operation along the entire route. The canal is over two-thirds of the entire distance of the trip from Buffalo to New York.

The application of electricity to the propulsion of canal boats is in no way limited to placing the motor directly on the boat, nor should it necessitate in any way changing or modifying the present barges. There are various other methods of applying this power; in fact, the most feasible and practical plan would seem to be some method in which the ordinary canal boat is propelled by an exterior motor, or a tug.

CHAPTER XI. .

METHODS OF APPLYING ELECTRICITY TO CANAL BOAT PROPULSION.—BOATS EQUIPPED WITH MOTORS.

55. Electrical canal boat propulsion is divisible under two distinct heads :—Propellers, in which the water forms part of the system of propulsion ; and hauling or towing systems. There are, however, five methods of applying these. They may be classified as follows :

A.—Propellers.

B.—Flexible submerged cable or chain towing.

C.—Rigid rail or rack haulage.

D.—Movable cable haulage.

E.—Motor locomotive haulage.

These five classes cover nearly all possible forms of practically operating canal boats by electric or other power.

Propellers.—Class A.

56. This class includes all forms of boat propellers operated by electric motor on the boat. Although the jet, paddle wheel and screw propellers are included, we shall only consider the screw. The other mechanisms have been found more or less impracticable for boat propulsion on canals.

As generally suggested, this method consists in placing an electric motor directly on the freight-carrying boat having at its stern a screw propeller which is revolved by the motor. The motor receives current from suitable contact wires suspended either over the canal waters or on the banks. As the boat has more or less lateral movement, the contact arrangement must be flexible, and as the canal (being fresh water) cannot well be used as a return, a double metallic circuit must be used. This will necessitate two wires for boats going in each direction. With any such system it will be found necessary to use some form of double overrunning trolley carriage as shown in the illustration (Fig. 50), and suitable switches and turnouts must be arranged in the contact wires.

The first boat so operated in this country was the "Frank W. Hawley," on the Erie Canal, near Rochester, during the latter part of 1893 (Figs. 51, 51a, 51b, 51c and 51d). She was named after the gentleman whose energy and enterprise led to the making of this noteworthy experiment the first of its kind in America.

This boat was an ordinary steam canal boat equipped

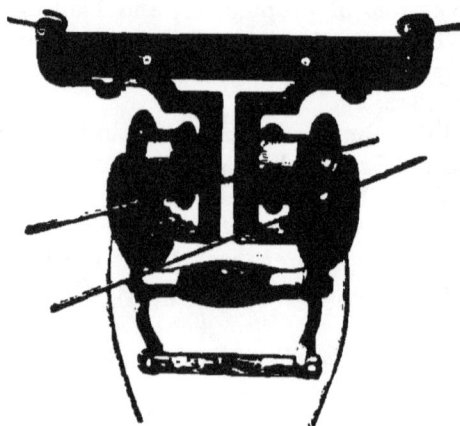

FIG. 50.—OVERHEAD DOUBLE TROLLEY.

with what is known as a dish-pan screw. The engine was disconnected from the shaft, and two Westinghouse electric motors of 25 h. p., street railway type, were substituted and directly connected to the screw. The motors received current from a pair of wires suspended over the canal through two ordinary underbearing trolley poles, as shown in the illustration (Fig. 51c). This arrangement was crude and caused a great deal of trouble as the lateral movement of the boat continually caused the trolley wheels to run off the wires. An arrangement of trolley carriage, somewhat similar to that shown in Fig. 50, was afterwards substi-

tuted.   The boat was run along the canal at quite a high
speed and its passage caused a great deal of excitement.

Various plans have been devised by which the boats are
permitted to move laterally without losing contact with
the trolley wire and by which two boats can pass each other

FIG. 51.—THE "FRANK W. HAWLEY." FIRST ELECTRICALLY PROPELLED CANAL
BOAT IN AMERICA—FIRST TRIP ON THE ERIE CANAL.

when going in the same direction.   Prominent among these
is a method devised by Mr. S. W. Gear, of Buffalo, which
consists in making the trolley contact and wire laterally
movable.   A general plan of this arrangement is shown in
the illustrations (Figs. 52 and 52a).   The trolley wire is

suspended on small carriages that are adapted to travel laterally on the span wire stretched across the canal. The

FIG. 51a.—PROPELLER OF THE "FRANK W. HAWLEY."

contact carriage on the boat is also laterally movable as shown. This makes apparently a complicated arrangement

FIG. 51b.—THE ELECTRICAL BOAT, "FRANK W. HAWLEY"— AT CANAL BANK.

for which there is really no necessity as the same results can be obtained with the simple method previously shown.

57. The placing of a motor directly or permanently on the freight-carrying boat is not the most feasible form of applying the propeller to canal boat propulsion. There are two others; (1) by the use of a false stern, rudder or portable motor which can be attached to the boat when desired and which contains the motor and screw, and (2) by the use of small propeller tugs.. The illustration (Fig. 53) shows quite a novel method devised by Mr. Samuel H. Jones, of Newark, N. J. The motor is placed in a box or casing and

FIG. 51c.—DECK OF THE "FRANK W. HAWLEY," SHOWING THE TROLLEY POLES.

revolves a propeller on the outside. This arrangement is either permanently or temporarily fastened to a rudder post, and the boat is steered by changing the position of the propeller and motor. A represents the rear end of the boat, B the casing, C the motor, D the screw, E the rudder post and tiller, and F are the wires going to the motor.

Various forms of applying a detachable motor and screw similar to the Jones plan can be devised. Another plan

FIG. 51d.—"FRANK W. HAWLEY," ELECTRIC CANAL BOAT, PASSING THROUGH LOCK.

is to have a false stern rigidly attached and holding the rudder, screw and motor.

58. The tug method can also be varied. The illustration (Fig. 54) shows a combination of the two which possesses

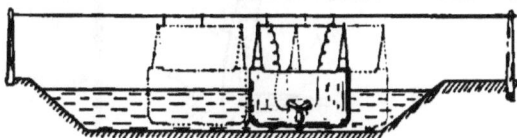

FIG. 52.—GEAR'S LATERALLY MOVABLE TROLLEY WIRE FOR CANAL BOATS.

several points of advantage over other methods of applying the screw. A represents an ordinary canal barge; B is a small boat about 25 feet long, 8 to 10 feet wide, and 5

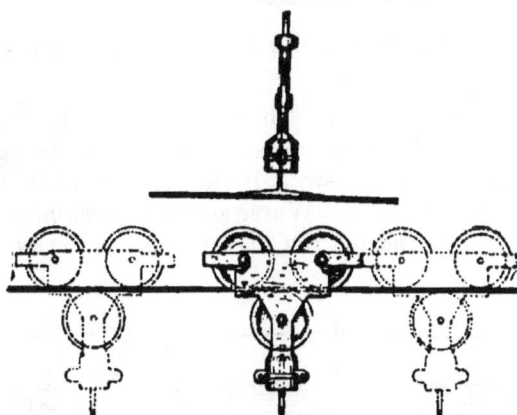

FIG. 52a.—LATERALLY MOVABLE TROLLEY WIRE AND HANGER.

feet draught, which holds the motor and screw, and which is rigidly attached to the freight-carrying boat. Contact is made with the trolley wire by the trolley carriage, C, which is connected with the boat by flexible connections. Such an arrangement can be applied to any number of boats, pushing one or two and towing one or more pairs, or the tug may simply tow boats.

Fig. 55 shows an arrangement of propelling mechanism for canal boats for which a patent has recently been issued

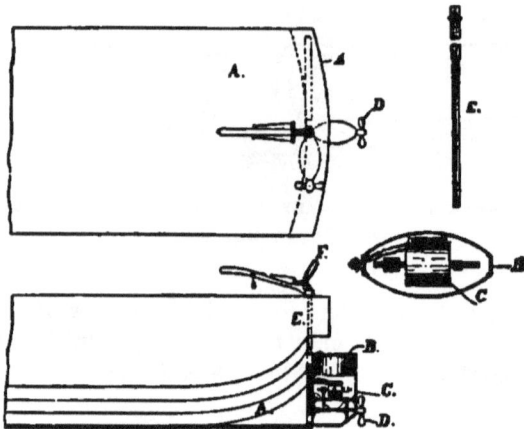

FIG. 53.—JONES' DETACHABLE PROPELLER AND RUDDER.

to B. C. Scott, of Belgium. The object of this device is to prevent the loss of power caused by the screw being but partly immersed. The screw, A, is enclosed in a chamber, B, at the stern of the boat. Water enters the chamber from the opening, C, at the bottom of the boat and is expelled

FIG. 54.—ELECTRIC TUG PROPELLER SYSTEM.

through the opening, D, at the stern. Gates or valves are located at both openings so that the chamber can be closed. When necessary the chamber can be exhausted through the valve, E. This will cause the water to rise in the chamber and immerse the screw completely when the inlet valve or gate is opened, and by opening the valve, D, the boat can be propelled with its screw completely immersed, without reference to the immersion of the boat. An electric motor is shown here revolving the propeller screw, and current is supplied by overhead wires as in the other propeller methods.

If the propeller is used it will readily be seen that some form of separate arrangement, either false stern or tug, must be adopted. With such an arrangement some of the requirements are met with, but it is doubtful whether the

Fig. 55.—Scott Combination of Jet and Propeller Operated by Electric Motor.

propeller is the most efficient and feasible means of canal boat propulsion.

The placing of a motor on the boat itself would cause but a very slight saving if any over the steamer. We could, perhaps, use a screw of small pitch turning at a high rate of speed, but the additional surface friction would make the total loss in the screw but very little, if any, less than with those at present in use; and as has been stated, the delivery of power at the motor shaft may be but very little if at all cheaper than as furnished at present by the engine. Such an arrangement, therefore, would really not be of any material advantage over the present steam propeller method.

59. Aside, however, from the equipment of the boat and the efficiency of the motor and propeller, the nature of the canal itself seems to be the principal objection to such a method of propulsion. It must be remembered that the canal is simply a large ditch, and no provision has been made in its structure to prevent the injurious action that a propeller would have on the walls and banks of the canal. The objection raised, when the steam propeller was first tried, was that the wash from the screw would ruin the banks of the canal. The screw agitates the water and this agitation or wash would, if there were enough of it, be very injurious to the structure of the walls. With the present steamers (and they are in the great minority) on the canals, at their low speed of between 2 and 2½ miles per hour, the banks of the canal are but slightly affected, but what would be the consequence if *all* boats on the canal were propelled in this way and at a *high* speed?

There can be no doubt, that a general adoption of high speed electric screw propellers upon the Erie Canal for all craft would cause great injury to the walls and banks. It would apparently be necessary to rebuild the present walls if the screw were universally resorted to, even though the present speed were not increased.

It would appear, therefore, that although a system of electric trolley propellers is perfectly practical and operative mechanically, its adoption, on the Erie Canal at least, would not cause any great improvement over the present steam propeller, as regards economy or so far as relates to the preservation of the canal structure.

Flexible Submerged Cable or Chain Towing.—Class B.

60. This system is extensively used on some of the canals and shallow canalized rivers throughout Europe. The steam engine has hitherto been used as a motor. Experiments have, however, been made recently with electricity.

The apparatus consists, in part, of a cable or chain laid at the bottom of the canal or waterway. It is lifted and passed over rollers or wheels on the boat, which firmly grip the chain or cable. These rollers are rotated by a motor on the boat and as they revolve they pull the boat along the waterway on the cable. The general arrangement of wires, motors, etc., would be very much the same as in the propeller method, but instead of turning a propeller, the motor would revolve the hauling drums or machinery.

Although it would be quite feasible to place the motor directly on each canal barge, it would be most advisable to employ separate towing boats ; not only for the room gained but for the other reasons previously given. The arrangement of distributing and contact wires and appliances would be about the same as in the previous class. Instead of having a double contact wire and double contact trolley carriage, the cable could be used as the return conductor. As this would necessitate only one contact wire, the contact arrangement and switching devices would be simple.

In France, Germany and Belgium, where the cable or

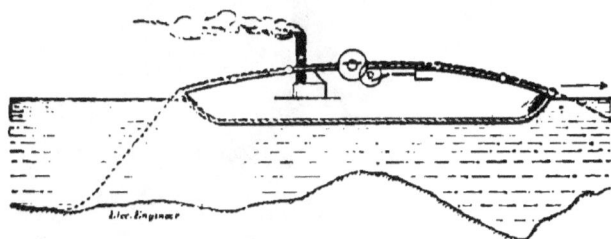

FIG. 56.—GERMAN METHOD OF SUBMERGED CABLE HAULAGE, SHOWING STEAMER.

chain towing method is used, it has been found necessary to employ a separate towing boat on which the boiler, engine and necessary hauling machinery is installed. This is necessary because the machinery used has been bulky, occupying a great deal of room ; and, the arrangement for taking up the chain or cable generally necessitates a separate boat. The European cable towing steamers are generally quite large and designed to take very heavy tows.

There are various methods of taking up and grasping the chain or cable. The illustration (Fig. 56) shows the plan used very extensively on the river Elbe, in Germany. In this method a flexible cable is used which passes over

Fig. 57.—Submerged Cable Haulage Formerly Employed on the Erie Canal.

the top of the boat and around the hauling drums that are revolved by a steam engine on the boat.

In 1873, a cable system of hauling was tried on the Erie Canal, but it did not prove successful. The method was similar to that in use in Belgium. The illustration (Fig. 57) shows the cable towing steamer "Gov. Clinton" which was operated on the Erie Canal. It will readily be seen from the cut how the cable is raised and passed over the central wheel or sheave which, by a peculiar arrangement, is so arranged as to grip it firmly. The boat was equipped with boiler and engine. Two such boats were operated for some time between 1873 and 1880, but they became so

FIG. 58.—CANAL BOAT "AMPERE," FRANCE.

objectionable and interfered with other boats to so great an extent that their use was discontinued.

There is great difficulty in properly gripping the chain or cable with any of these methods, and various plans have been devised for obtaining a good hold between the hauling drum or wheel and the chain or cable. Among these is a plan devised by a French inventor in which a sprocket wheel is substituted for the friction drum or pulley and the chain used is so arranged as to engage with the sprocket wheel. This plan is more or less impractical as the chain very soon gets out of adjustment.

In France, on the Seine, a peculiar and novel system of chain towing has been in use for some time. Although electricity does not furnish the actual propelling power, it

plays quite an important part in the general method. A
chain is used instead of a flexible cable, and in order to
obtain the necessary grip upon the chain, the drum over
which it passes is mag-
netized by an electric
current and the neces-
sary grip obtained by
magnetic adhesion.
(Fig. 58.) In the illus-
tration (Fig. 58a) there
are shown transverse
and longitudinal sec-
tions and a plan of the
towing steamer. This
method was devised by
M. De Bovet and con-
structed by the Com-
pagnie de Touage de la
Basse Seine et l'Oise.
The boat called the
"Ampere," can be pro-
pelled either by haul-
ing on the chain or by
means of the propeller
at the stern as will be
seen. Either or both
can be used at once.
The engines are of 150
horse-power. The chain
passes over the bow,
around the towing pul-
ley A, and is guided by
a non-magnetic guide
pulley, B, and passes out
over guide pulley C.
The latter is very mas-
sive, and is made of iron

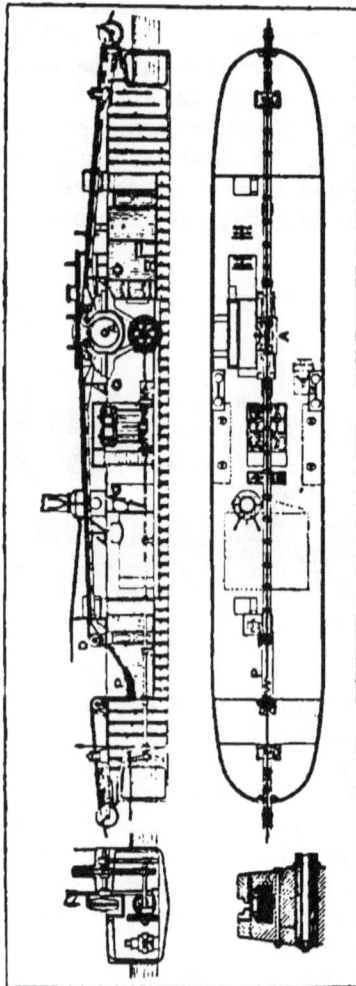

Fig. 58a.—Magnetic Adhesion Towing Steamer on the Seine, France.

for the reason that if at that point the guide wheel is put
in contact it gives to the magnetic lines of force an easier
passage than that offered by the chain, and the latter no
longer serving to close the magnetic circuit, releases itself

more easily under the action of the weak tension of the
slack towards the rear. A finger of non-metallic metal is
arranged above the pulley A, so as to make sure of the
disengagement of the chain.

In this method of towing it is sometimes necessary to be
able to control the paying out of the chain at the stern of
the vessel, and for that purpose, after passing over the
grip pulley A, part of the chain is deposited loosely in the
pit P, provided for that purpose ; keeping it there when
there is sufficient slack and paying it out again when more
slack is needed. The paying-out chain at the rear of the
vessel is, therefore, also magnetized like the towing pulley,
but not so strongly, as the braking effort is far less than
is required for traction.

When the chain has little slack, the action of the pulley
C is insufficient to produce the disengagement of the chain
from the main pulley A, and then by magnetizing the
pulley D, it assists the pulley C in pulling off the chain
from the main driving pulley.

The construction of the magnetic pulley is shown in section. The magnetic friction being a function of the
intensity of the current, it is evident that the frictional
force can be regulated to a nicety, and sudden strains put
upon the chain merely allow it to slip a trifle without
causing rupture.

61. The substitution of the electric motor for the steam
engine and boiler, as used in the European methods, would
without question be perfectly feasible, and the economizing
of space effected might make it possible to locate the
towing apparatus directly on the canal boat, if such were
desirable. The illustrations show a method of applying
such towing apparatus directly to canal boats. Fig. 59
shows a plan of this method on a canal. Fig. 59a is a view
of the motor on the boat with the hauling machinery.

France possesses a large number of canals which render
water communication very easy between different parts of
the country. Unfortunately, up to the present time, the
hauling has been done most generally by men or animals.
The time occupied was frequently very long, and, as a
result, railway transportation has been preferred. It was,

therefore, necessary to find a remedy for this backward
state of affairs. The first investigation with this end in
view goes back to the year 1888, when M. Maurice Lévy
made a number of experiments with cable hauling in the
neighborhood of Paris, at the junction of the St. Maur and
St. Maurice canals. The system was based on the employ-

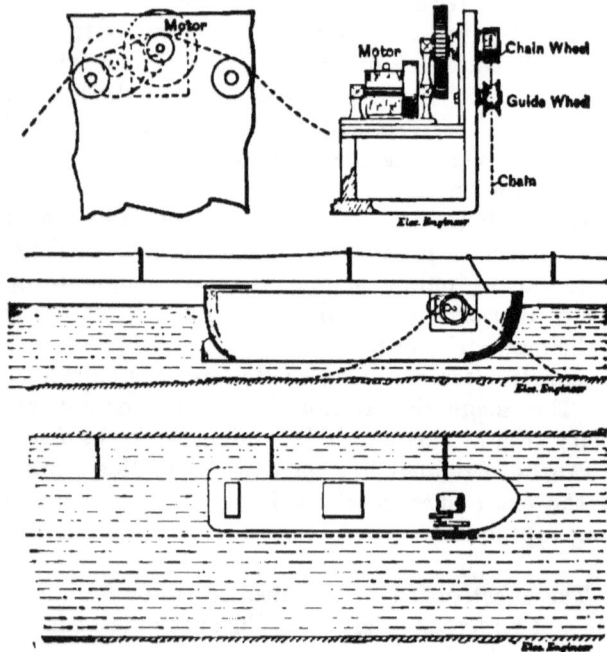

FIGS. 59 AND 59a.—ELECTRIC MOTOR SUBMERGED CABLE
TOWING SYSTEM.

ment of an endless wire cable on each bank, carried on supports provided with pulleys. The cable was operated by a
fixed motor, and to it the boats were attached by means of
hauling ropes. This system gave a speed of four kilo-
meters per hour. It constituted a considerable improvement on the existing methods but did not give entire
satisfaction. Further reference will be made to it.

In 1890 there was talk of employing electric traction,
and comparative plans were drawn up for steam and

electricity. The first plans for steam were rejected as being too costly and impracticable; as for electricity one could utilize as motor power the water falls, found on the route of the canals themselves. This idea was worked at for several years and finally, towards the end of 1893, an installation of the submerged flexible cable type (Fig. 60) was put in on the Bourgogne Canal, which unites the river Yonne and the river Saone. The Yonne being an affluent of the Seine, this canal establishes communication between the English Channel and the Mediterranean. On this canal the hauling machinery is now operated by an

FIG. 60.—ELECTRICAL CANAL BOAT, BOURGOGNE CANAL, FRANCE. (SUBMERGED FLEXIBLE CABLE HAULAGE.)

electric motor on the boat, receiving current from an overhead trolley circuit.

The installation has been established by the Department of Bridges and Roads for a length of six kilometres close to an underground tunnel 3.3 kilometres long. The motive power is obtained from water falls having a fall of 7.5 metres in two successive locks. There are thus two generating stations established at the two ends of this junction canal. In each the turbine drives a Gramme dynamo, one running at 1,200 revolutions per minute and 380 volts, and the other at 900 revolutions and 270 volts. Each of the dynamos is appropriate to the conditions existing at the locks and the corresponding turbine. The dynamos are connected after the manner shown in the accompanying diagram (Fig. 60a). A and B are the two dynamos connected to a bronze overhead feeder of 8 to 10 mm. diameter, held by porcelain insulators mounted on poles. The dynamos are coupled in series. D D are the trolley wires, and E the towing motor which drives the chain drum.

A storage battery of 250 cells, having a capacity of 150 ampere hours and a discharging current of 15 amperes, is coupled in parallel. These are Chloride batteries made by the Société pour le Travail Electrique des Métaux. The

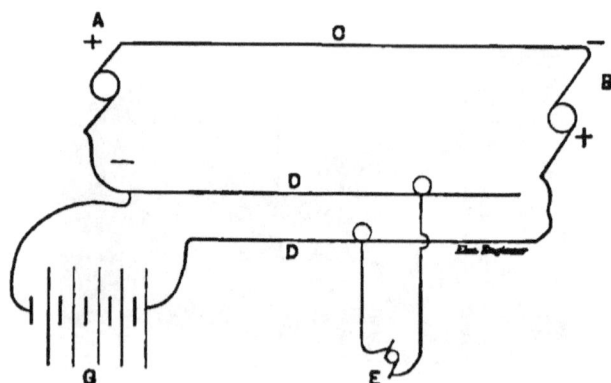

FIG. 60a.—DIAGRAM OF CIRCUITS OF FRENCH ELECTRIC CANAL TOWING.

mean rate of discharge varies, according to the load, from 10 to 25 amperes at the potential above given. The stretch of six kilometres is covered in less than an hour.

Special arrangements have been made for the automatic regulation of the turbine. The water gates are opened or closed according to the power furnished by the machine. The tunnel of 3.3 kilometres mentioned is lighted by incandescent lamps which are branched in multiple on the power circuit. It may be said that this electric canal installation has given such satisfaction that it is certain that before long similar methods will be largely employed in France, where canal locks are numerous. It is the first to be operated on a practical, commercial basis.

62. However, the preferable way to operate any electric propelling system, would appear to be in having no machinery on the freight carrying canal barge. Some form of separate towing or hauling boat or some arrangement of motor detachable from the canal boat would be more feasible than placing the motor directly on the boat.

The first plan would correspond to the European methods, with the exception that the towing or hauling boat would be comparatively small. The second plan would seemingly embody more or less complication and is, perhaps, not as practical as the first. The illustration (Fig. 61) shows a

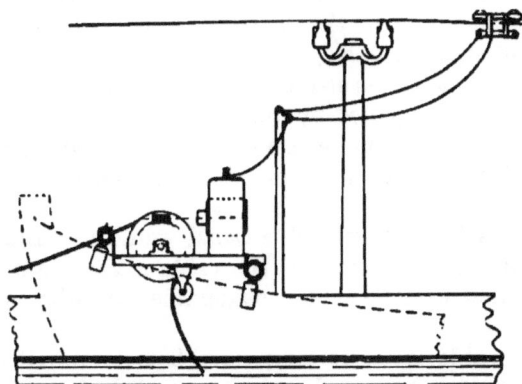

Fig. 61.—BUSSER METHOD OF SUBMERGED FLEXIBLE CABLE
HAULAGE.

plan devised by Mr. Otto Büsser, a German engineer, in which a detachable prow is used holding the motor and the hauling machinery. The general operation is similar to that in other cable towing methods with the exception that the motor, instead of being fixed on the canal boat, is placed with the hauling machinery on a platform which can be attached to the gunwales of the boat. Current is fed by a trolley.

Although a system of electric cable or chain towing would be more efficient than the propeller, and free from some of its objections and would hardly cost very much more to install, it has various difficulties which make its practical adoption, on the Erie Canal, for example, quite doubtful. Such systems would, however, be most satisfactory on comparatively short, level and shallow waterways, and where a very large number of boats could be hauled by one towing boat. With regard, however, to the Erie it may suffice to say that it has been tried and found wanting.

Amongst the objections to the application of a method of cable or chain towing or haulage by electric motor, various troubles may be mentioned:

The first is the difficulty in rounding curves and in steering the boat. As the chain or cable lies loosely at the bottom it frequently comes very close to the banks, and as the boat must follow the chain or cable, it runs into the banks and sustains damage.

The boats cannot be steered, and frequently cause injury to other boats propelled by mules.

If two cables or chains are used, one for each direction, it is difficult to keep them from interfering with one another.

The many locks on the Erie Canal would necessitate the taking up and dropping of the cable a great number of times, which would cause considerable difficulty and loss of time.

As the cable must be long enough to be raised by the different towing boats, a large amount of slack is necessary which must be pulled in by each boat. Slipping of the cable on the hauling drum or sheave would also be experienced in operating heavy tows.

Rigid Rail or Rack Haulage.—Class C.

63. This method consists in placing a rigid rail or rack parallel to the canal. The rail or rack may either be placed on the banks or over the canal or may be submerged. Engaging the rail or rack are rollers or pinions, extending out from the boat on suitable arms or supports, connected with a motor on the boat by which they are rotated.

The boat is propelled by the rotation of these rollers or pinions on the rail or rack by the motor on the boat. The

motor receives current from a similar contact arrangement
as in the previous system (one wire only being used each
way).   If the exterior rail is used, the wires can be
supported by the structure holding the rail.

This plan has never been practically tried, although
various methods have been devised of operating in this
manner with a rail or rack exterior to the canal and an
engine and boiler on the boat.   A plan devised by Mr.
N. P. Otis, of Yonkers, in which a rack and pinion is used,
is shown in Fig. 62.   The rack A is placed along the bank
of the canal.   On the canal boat, B, is placed an engine
which revolves the pinion, C, at the side of the boat and
which engages with the rack.   At the other end of the

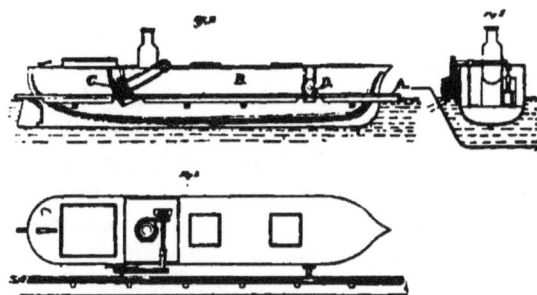

FIG. 62.—OTIS METHOD OF PROPELLING CANAL BOATS.

boat is another pinion, D, engaging the rack, but not
rotated, to keep the boat in line.   Both of these pinions
slide in guides to allow for the varying draught of the
boat.   It will readily be seen that by rotating the pinion C
by the engine, a direct pull is obtained on the rack, and
the boat pulled along the waterway as the pinion revolves.
It would be very simple to substitute an electric motor for
the engine.

64. Instead of placing the electric motor directly on the
barge, it may be placed on a separate towing or hauling
boat as in either of the previous methods.   A novel and
original plan of rail haulage is shown in Fig. 63.   The rail
is suspended over the canal and a separate hauling boat is

used. The cut shows a transverse section of the canal with the towing boat only. Suspended from the cable, A, strung across the canal, are the double headed rails, one for each direction. The contact wire, C, is suspended below the rail and supported by insulators fastened to the rail which forms the return. On the boat, D (which is connected to

Fig. 63.—Overhead Rigid Rail Haulage.

the barges), is placed an electric motor receiving current through the contact arm, E. The rollers, F, engage each side of the rail and are supported as shown. As the motor revolves the rollers, the boat is pulled along by the traction of the rollers on the rail. Arrangements must be made to give the rollers the necessary pressure on the rail, and some means is necessary for readily disconnecting the boat from the rail. Instead of using rails, a rack and pinion can be used. The hauling boat may also be equipped

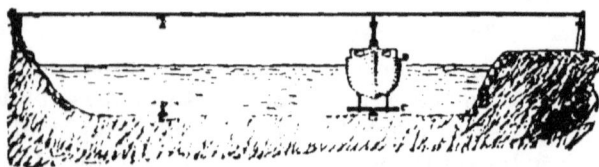

Fig. 64.—Submerged Rigid Rail Boat Haulage.

with a screw so that it may be propelled through locks. A different contact arrangement is necessary if this is done.

65. The placing of the rail or rack at the bottom of the canal and operating, in a manner similar to the previous, is also a somewhat new and novel suggestion. The illustration (Fig. 64) shows such a method, which is practically

the reverse of the one last described.   The cut shows a transverse section of a canal with the hauling boat.   Over the canal are suspended two trolley wires, A, one for each direction, but they may also be supported on posts along the banks.   The boat, B, is equipped with an electric motor receiving current through a contact arrangement as shown. This motor revolves the two rollers, C, below the bottom of the boat.   The rail, D, is anchored at the bottom of the canal, and the rollers engage with it and propel the boat.

Fig. 65 shows a general arrangement of the circuit connections and propelling mechanism that could be used

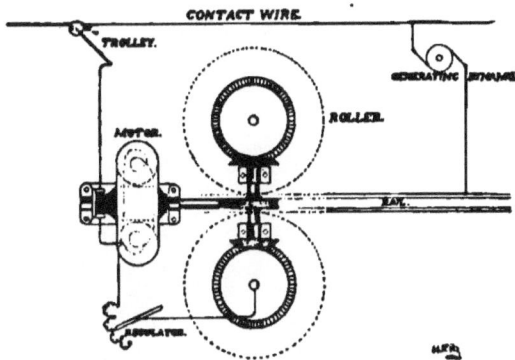

FIG. 65.—ELECTRIC MOTOR, GEARING AND FRICTION DRUMS
USED WITH RIGID RAIL.

in either of the preceding methods of rigid rail or rack propulsion.

66. As most canals have slope walls, some form of overhead construction would be used in rack or rail methods unless the walls were changed ; or it might be possible to have arms extending out from the hauling boat to the rack or rail on the bank.

In each of the classes already described it is necessary to place upon the propelling boat more or less stationary or detachable machinery.   It is true that this may be a separate towing boat carrying no freight, but such boats

would interfere to some degree with the present methods
of locking and, if there were a large number, would cause
more or less trouble. Over 95 per cent. of the present
barges are towed by mule power from the banks. There is
no objection to this method were it not for the large cost of
operation. Such a plan has various advantages which will
readily be seen.

In the other methods to be described, this plan is
adhered to. The present canal barges are left as they are
and are propelled by power exterior to the waterway, as
in the case of mule haulage.

CHAPTER XII.

METHODS OF ELECTRIC CANAL BOAT PROPULSION WITH MOTOR EXTERIOR TO BOAT.

Movable Cable Haulage.—Class D.

67. This system, although not very practical or feasible on the Erie, could be successfully applied on some other canals to boat propulsion. The propelling power is furnished by a moving cable operated parallel to the canal. Extending from the boat, are suitable arms, to which devices are attached, adapted to grip and hold the moving cable. The boat is pulled along the waterway by the movement of the cable, and can be started or stopped, by connecting or disconnecting the grip with the cable by suitable devices on the boat. The boat is therefore operated mechanically. Messrs. Oriole and Maurice Levy, in 1888, proposed and experimented with this method on some of the French canals, as already stated. The cable stations were, however, operated by steam.

The first experiments in this direction, Figs. 66, 67 and 68, were undertaken at the junction of the Saint Maur and Saint Maurice canals. This point was especially selected because the canals meet at right angles and hence present peculiar difficulties in turning the boats. An endless metallic cable is installed upon each bank, a few yards from the edge, in order to leave the tow path free. It is supported here and there by channeled pulleys, which are loose upon metallic supports, from six to ten feet in height. On a straight line, these pulleys are vertical, and are ten inches in diameter. On curves they are more or less inclined and have a diameter of five feet. At its starting point, the cable passes over three large pulleys, actuated by a steam engine, placed in a small power house at the edge of the canal. To the right, there is a fourth pulley carried by a small car provided with a counterpoise which serves to keep the cable uniformly taut. The cable is provided with links to which is affixed the rope that hauls the

FIG. 66.—LEVY SYSTEM OF MOVABLE CABLE HAULAGE.

boat along. These links, fixed between rings, are capable of revolving freely upon the cable so as to avoid troubles from the latter's tension. In order to prevent the cable from jumping out of the pulley channels, it is kept in place by a small overriding roller, and the flanges of the pulleys are notched, so as to allow the hauling links to go by. The cable runs at 2½ miles an hour.

The system herein proposed is, however, electrical, as the cable lengths are operated by large electric motors located along the line of the canal and receive current from one or two very large generating stations located at a cheap source of power such as a water-fall. No distributing or contact wires are used along the canal excepting the

FIG. 67.

power wires from the large central generating station to the various electric cable stations. Fig. 69 shows a general arrangement of this method. As will be seen, electricity is generated by means of large alternating dynamos, which are operated by water-power. The current is transmitted, at high pressure, to the various cable operating plants along the canal where it is reduced in pressure and synchronous or induction motors are operated. The motors move the cable machinery, which is similar to that of any cable railway plant. The cable would be operated in lengths of about 5 miles, each way, from the power station, the outgoing cable being used for boats in one direction and the

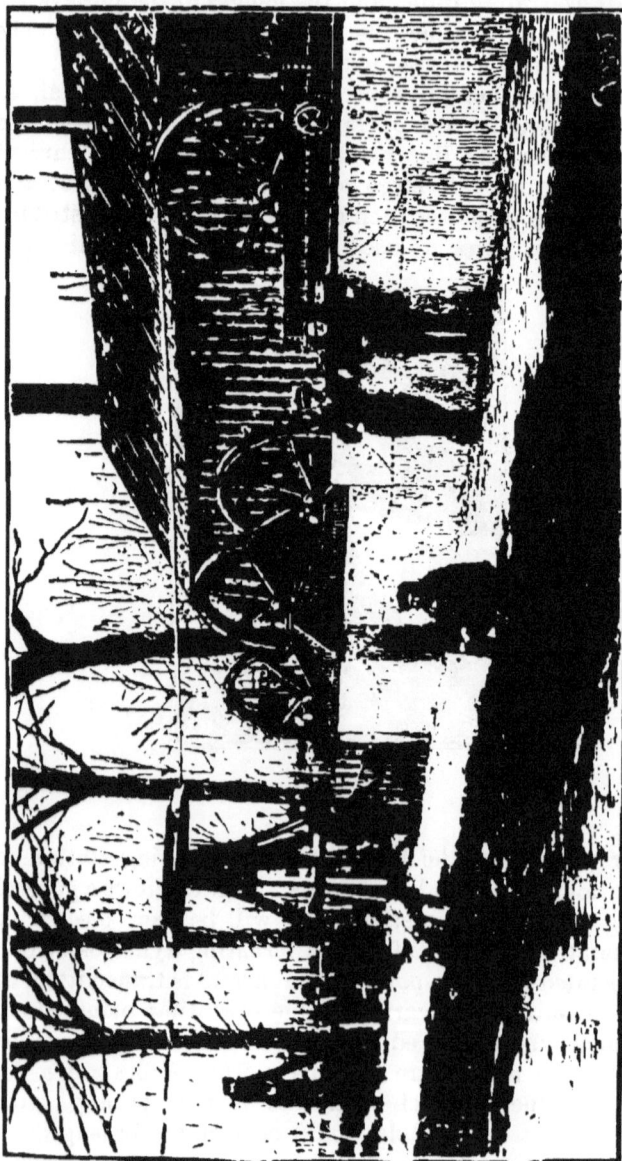

Fig. 68.—Levy System of Movable Cable Canal Boat Haulage.—Power House.

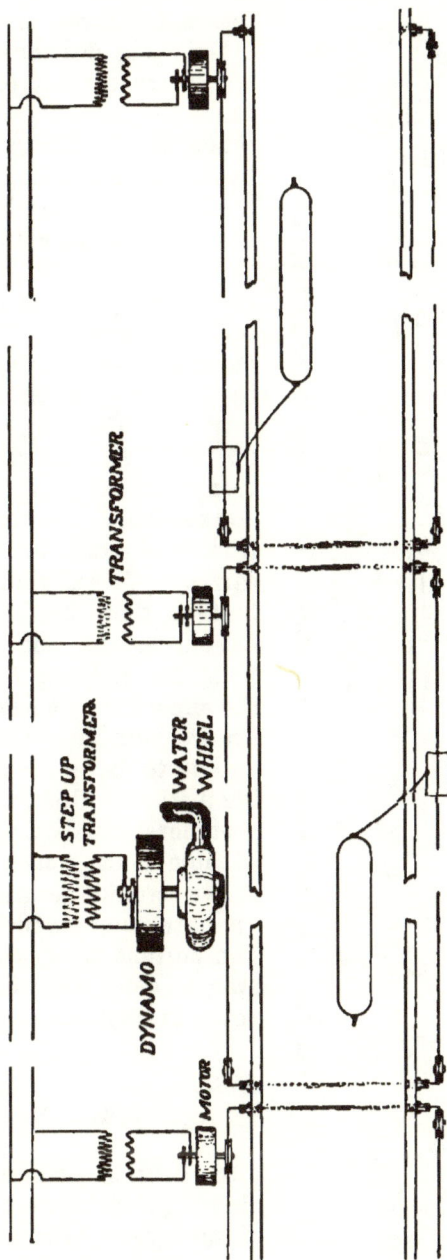

FIG. 69.—CIRCUITS AND GENERAL POWER PLAN OF MOVABLE CABLE HAULAGE SYSTEM WITH ELECTRIC POWER HOUSE.

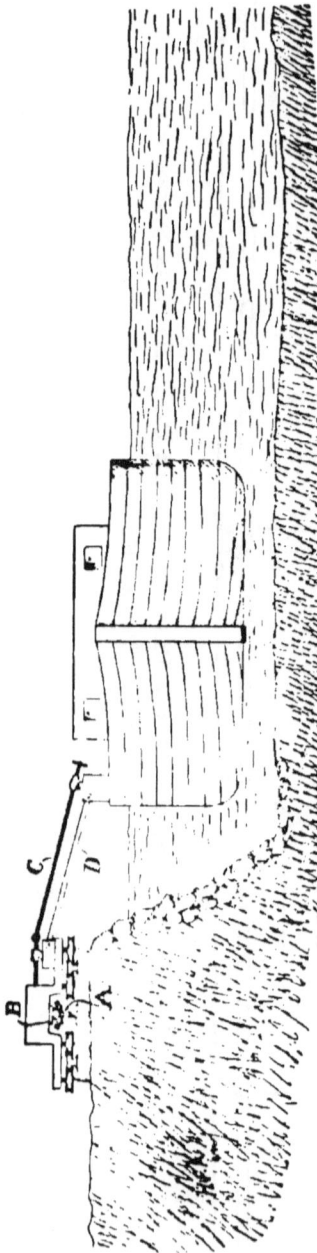

FIG. 69a.—SECTION OF CANAL WITH MOVABLE CABLE AND ELECTRIC MOTOR.

return for those in the other. Fig. 69a shows a transverse section of a canal equipped with this method. Along the bank of the canal a track is laid, upon which the grip carriage runs, which is suitably held to the rails, by the rollers, as shown. The moving cable is grasped by the grip, B, on the carriage. The grip, B, is operated by a connection, C, from the boat. A rigid connection, D, extends from the boat to the carriage. As will readily be seen, there are some difficulties in such a plan, but judging from the experiments made in France, this plan would appear to be a promising method. Upon very short level canals, a plan of cable haulage such as shown would certainly operate successfully. The current must, however, be generated where power is very cheap.

Motor Locomotive Haulage.—Class E.

68. In this plan, the old and primitive method of haulage or towing as by mules is adhered to;

but, for the slow and uneconomical animal, an efficient electro-mechanical mule is substituted. The boats are not equipped with electric machinery, but any ordinary boat can be towed or hauled without any previous preparation.

Such systems of operating canal boats were among the first proposed. In the earlier forms, the steam locomotive was used instead of the electric motor. This was run upon a track on the banks of the canal and towed the boats connected therewith. For the steam engine an electric locomotive can very readily be substituted and current can be supplied in any suitable way. There would practically

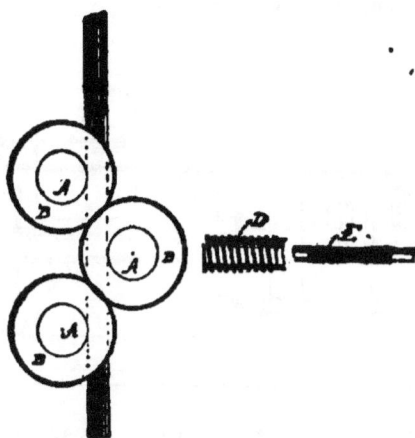

FIG. 70.—DAVIS' ELECTRIC CANAL BOAT TOWING SYSTEM.

be an electric railway along the banks; the boats being "trailers," connected by tow line to the motor cars.

69. Among the plans that have been devised for operating electric locomotives on suitable tracks along the banks of the canal and towing the boats in various fashion, is the method of Mr. T. D. Davis, of Syracuse, N. Y., as shown in the illustration (Figs. 70, 70a, 70b, 70c, 70d). The plan contemplates the laying a narrow gauge rail-track on each bank of the canal and moving the boats in trains or tows of four or six boats each by means of a small car furnished

with a device for gripping the rail, to be driven by an electric motor from an overhead trolley line. The boats are connected with the inner rail, laid near the water's edge, and require no steering. In some of the larger cities on the line of the canal where it might be impossible to lay a track on the berm bank a double track may be laid on the tow-path and the outside tows steered as they now are.

Figs. 70 and 70a show the principle of the device for towing the boats. AAA are three horizontal wheels secured to vertical shafts carrying on their upper ends three cog wheels, BBB, by which they are geared together, and all have a uniform speed. Two of the wheels, A, are on one side of the rail. The third wheel is placed in a movable frame on the

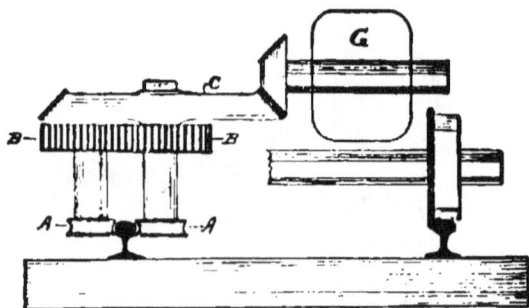

FIG. 70a.—DAVIS' ELECTRIC CANAL BOAT TOWING SYSTEM.

opposite side and is forced against the rail by the spring, D, surrounding a shaft, on one end of which is a threaded extension, E, connected by gearing with a hand wheel within easy reach of the motorman. The object of this is to secure an elastic grip of any desired pressure of the wheels on the rail. The shaft of one of the wheels, A, extends above the cog wheel, B, and is provided with a bevel wheel, C, which is driven by the motor, G; the whole carried on a small platform car and driven by a trolley connection. This device insures a positive grip on the rail and utilizes the entire amount of power received from the motor, with the exception of a small amount of loss by unavoidable friction.

Fig. 70b shows a light triangular wooden frame. This is provided at its wide end with a pair of T-hinges by which it is connected to the side of a boat as shown at Fig. 70c. A pair of slotted irons about four feet long, secured to the side of the boat, have their upper end open for the reception of the T-hinges which are secured at any desired height by pins. This frame is provided with an iron bar, B, moving freely in the frame and provided near its outer end with three small wheels; one vertical wheel, C, rolling on the top of the rail, E, carrying the weight of the free portion of the frame, and the two horizontal wheels, DD, rolling, one on each side of the rail.

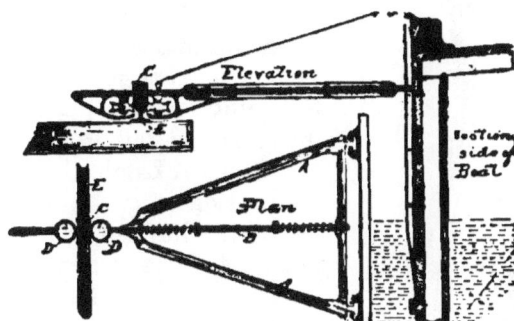

FIGS. 70b AND 70c.—DAVIS' ELECTRIC CANAL BOAT TOWING SYSTEM.

The bar, B, is provided with two spiral springs as shown and also with an extension or guard at its extreme end. The spring will compensate for any sudden jar or movement of the boat, thus relieving the boat and rail and securing an elastic, instead of a rigid, connection.

When a boat enters a lock, the free end of the frame, A, is raised by a lanyard and turned over on the deck, as shown by the boat leaving the lock at the left of Fig. 70d.

It will readily be seen that an ordinary electric hauling locomotive, as used in mining or factory work, with the necessary trolley and contact wire, and so arranged as to resist the side pull of the boats, could be used in the Davis system. There have also been various other suggestions similar to this, nearly all necessitating the use of tracks on one or both banks of the canal (preferably both) on which runs an electric hauling locomotive, which is suitably held to the track, either by its own weight or by peculiarly constructed track, and is operated by a person riding on the hauling locomotive. Perhaps a very feasible form of this method of hauling would be to erect along the banks of the canal, a form of telpher system with suitable devices for holding the car to the cable or rails.

An ingenious form of such a system as that referred to in the foregoing paragraph, has been invented and put in operation by Mr. Richard Lamb, C. E.; and our illustration (Fig. 71) shows it in use on the Raritan Canal at Trenton, N. J. This system is intended, also, for the haulage of logs in lumber camps, to which it has been applied with considerable success.

As shown in Figs. 71 and 71a the system consists of a cableway along the towpath supported by brackets from 100 to 150 feet apart, placed upon trees or substantial columns and insulated from the ground. The main, or bearing cable has a copper core, having large steel wires surrounding it, giving ample strength as well as being a good conductor for the current, and presenting a good wearing surface for truck wheels to roll upon.

The cable rests in steel saddles on the bracket, having wedge clamps upon them to hold the cable rigidly to the

FIG. 70d.—DAVIS' ELECTRIC CANAL BOAT TOWING SYSTEM—BOATS LEAVING AND ENTERING LOCK.

Fig. 71.—The Lamb Electrical Cableway for Canal Boat Towing and Logging.

bracket; these saddles are insulated from the brackets. Under the bracket proper is an auxiliary bracket having a V-shaped saddle with its mouth opened at the top, and having jaws projecting inwardly to engage the ½-inch traction cable, preventing it from coming out of the saddle, except when temporarily lifted when the car passes a bracket. Owing to this feature, the cable is not confined to a straight line, but can be located with right and left curves. The bracket is easily removed and in canal-boat towing a simple permanent bracket is used. This ½-inch cable is attached firmly at either end and grounded. In practice this cable does not require to be very taut.

The motor is made with a truck, having grooved wheels to run upon the cable. A horizontal axle on the frame of the truck, located below the centre and between the wheels, supports a hanging-frame. An elliptically grooved sheave is attached to this frame, which is revolved by means of a newly patented worm, or wedge-gearing, driven by a 5 K. W. Lundell electric motor, with vertical shaft.

By taking a couple of turns of the ½-inch traction cable, around the elliptically grooved sheave, when the electric motor revolves the gearing, the sheave winds up, and at the same time pays out on the ½-inch cable, the action being the same as that of any elevator, only the direction of the pull of this motor is horizontal instead of vertical. The direction of the pull on the traction cable is parallel with the bearing cable; in consequence, the motor can easily climb steep grades.

The hanging-frame is divided just below the level of the cable, and is bolted together, with provision for thorough insulation of the upper from the lower part of the hanging frame. An insulated wire is connected with the upper part of the frame; this extends to the rheostat, which in turn is connected with the motor. The return current is allowed to pass through the frame to the elliptically-grooved sheave, then through the ½-inch cable to the ground.

In canal-boat service the towing hawser contains insulated wires. This cable is rigidly attached to an eye-bolt, just below the point of insulation on the hanging-frame, and from the end of the rope, the various wires are connected with their respective connections. The bight of the rope is connected with a clamp made of non-conducting material; the socket of this clamp, and the pin which engages in it, which contains the wires leading to the reversing switch and rheostat, are made irregular in shape, in such wise that corresponding wires are obliged to come in contact, when the clamp is connected.

In operating, a canal boat will apply for a motor; if the boat does not own a rheostat one will be placed on board.

FIG. 71a.—THE LAMB ELECTRIC CABLEWAY.

The towing rope will be attached to the samson-post, leaving the end of the rope free. The wires in the cable will be connected with their respective wires in the rheostat by the clamp, as described, and the boat proceeds.

On approaching another boat coming in the opposite direction, motors are stopped, cables are disconnected, and boats exchange cables and, consequently, motors, and proceed. Of course, an extra cable, one above the other, or one on both sides of the banks, would obviate the necessity of exchanging motors.

The trial plant illustrated, is operated with a 15 K. W. Edison dynamo, at 220 volts. The motor is a 5 K. W. Lundell, provided with a metal bonnet to protect it from rain, but the bonnet is not shown in the illustration. The ease with which heavy logs can be pulled in from a distance on either side of the cable, and lifted and carried upon the car, is remarkable. The large scow shown is put in motion quickly; boats can be pulled at the rate of six miles per hour with the motor now made.

70. The method of hauling canal boats in trains by a large locomotive running along the banks would perhaps be somewhat objectionable on the Erie Canal. The track —either surface or elevated—and the hauling locomotives would be quite large, and would interfere with the present methods of propelling by horses. Such an arrangement would also take up both banks of the canal and would be costly to install.

There would seem to be a number of objections to placing a track upon the banks of a canal and hauling the boats in trains by large locomotives running on the track. There is, however, no particular reason for adhering strictly to this plan. Instead of operating large hauling locomotives in the manner described, a small motor could be supported by a light structure running parallel to the canal and one or two boats could be pulled by each motor. As the motorman would be dispensed with in such an arrangement, provision must be made to regulate the motor in some other manner. The supporting structure should be light and strong and not interfere with the present methods of propulsion.

Such a plan has been devised and proposed by Mr. Joseph Sachs. It contemplates erecting a suitable structure along the canal and supporting a rail or similar device thereon. This structure is arranged so as to permit a continuance of the present methods. The hauling motor runs upon the rail so that it cannot be displaced and is connected to the boat by a tow line or otherwise. The distinctive features of this plan consist in regulating the hauling motor from the boat to which it is connected and preferably pulling each boat by a single motor, whether the boats are connected or not. This admits of

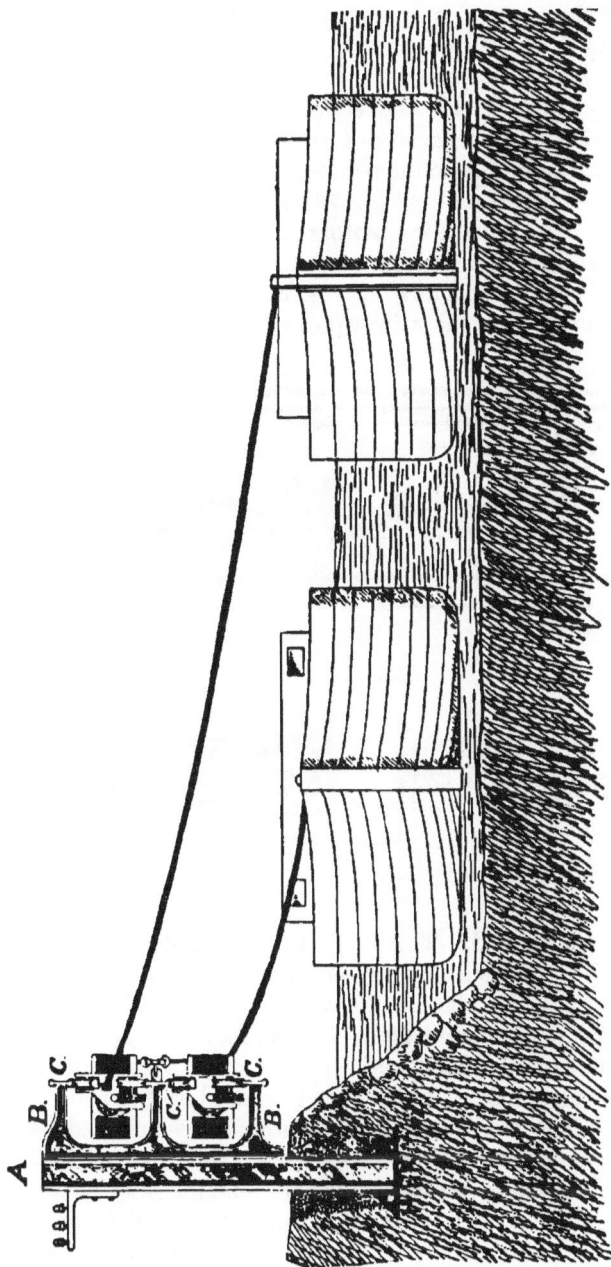

Fig. 72.—Sachs' System of Electric Motor Haulage; Duplex Structure; Regulation from Canal Boat.

making the motors comparatively small, and of erecting a small, light structure.

The current is supplied by wires supported on the structure which also forms the return circuit. It will readily appear that a great variety of supporting rails and structures for the motor could be devised, and several different forms are illustrated. Fig. 72 shows a type of duplex structure which permits motors going in opposite directions to run upon a structure on only one side of the canal. Ranging along one bank of the canal is the supporting structure to which the rails, upon which the hauling motor

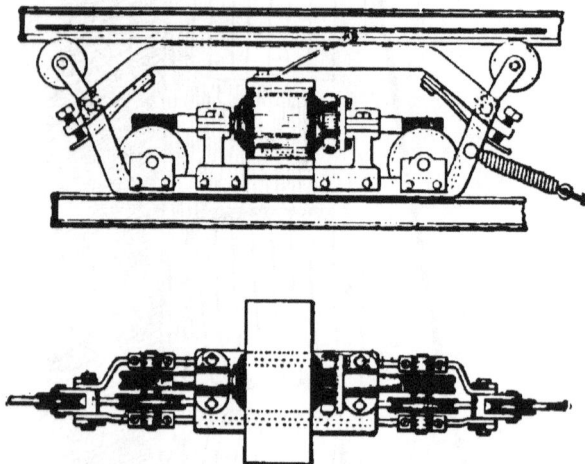

FIGS. 72a AND 72b.—"BICYCLE" MOTOR HAULER.

runs, are attached. This structure consists of iron or wooden posts, A, standing about 8 to 10 feet above ground and very solidly planted to a depth of at least 6 feet. These posts are erected from 15 to 30 feet apart, as may be found necessary. Fastened to the upper portion of each post are the yokes, B, which face the canal and to the three arms of which the double headed rails, C, are fastened. On the opposite side of the supporting posts or pillars are suitable arms and insulators for various wires, and fastened to the central rail is the contact rail which furnishes current to the motors above or below. These motors or haulers, it will be seen, are small "bicycle" electric locomotives which are run between the middle and upper and middle and lower rails. Traction is obtained by pressing the lower flanged rollers or wheels of the hauler against the lower rail upon which the motor runs. This is accomplished by causing the upper or pressure roller to exert an upward pressure, thereby forcing the lower traction wheels into intimate contact with the lower rail. The upper and lower motors travel in opposite directions and are regulated from the boats to which they are connected by an ordinary tow line.

It is unnecessary to go into the detailed construction of this particular plan. It will readily appear that this struc-

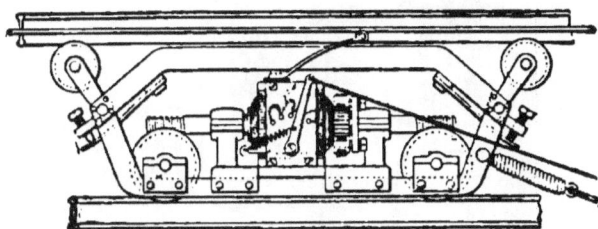

FIG. 72c.—"BICYCLE" MOTOR HAULAGE, WITH REGULATOR ON CARRIAGE.

ture can be suitably constructed and erected along a canal. A view of the motor and carriage is given in Figs. 72a and 72b.

The current is taken by the trolley arm from the supply rail and is brought to one terminal of the motor, the other being connected with the rheostat, which is in this case placed on the boat; and the other wire from the rheostat goes to the frame and to the rails which are used as a return. A reversing switch may also be placed on the boat. In fact any ordinary motor regulator can be used, it being merely a question of putting the necessary wires into the cable connecting the motor with the regulator. It may, however, be found preferable to leave the regulator upon the motor. In fact, if any form of series parallel controller is used, it would be somewhat troublesome to run a cable containing a large number of wires from the motor to the boat. If such should be used, the regulator is placed as shown in the illustration, and the handle connected by a cord or chain to the boat (Fig. 72c).

71. The illustration (Fig. 73) shows another form of duplex construction. In this form there are two rails, side by side, used for each motor, and the wheels of the motor are so arranged as to clamp or grip these rails; the upper being the driving wheels. The general idea of the previous plan, however, is adhered to, viz.: duplex struc-

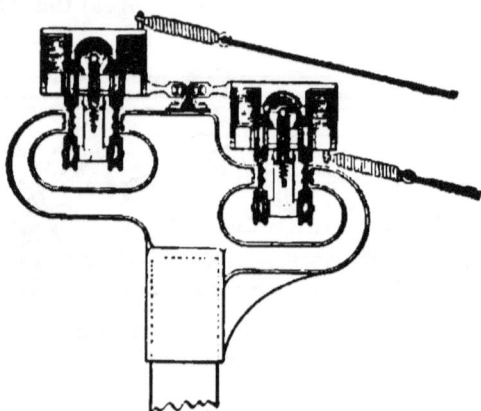

Fig. 73.—Double Rail Duplex Structure.

ture on one bank of the canal, single contact rail for both motors, regulation from boat and structure or ground return.

72. By the duplex structure arrangement, as shown, only one bank of the canal is used and the other is left free. As the towing path on the Erie crosses from one bank to the other, such an arrangement would meet with some difficulties. Instead of operating motors to run in both directions in this manner, a rail or support could be placed on each side of the canal on suitable elevated supports, so as to clear the towing path, and several such methods of

construction are shown in the illustrations (Figs. 74 and
74a). The general operation and regulation would be the
same; and the plan in Fig. 74b shows the arrangement
common to all these methods.

We can, however, also support the rail upon which the
motor runs, over the canal. Such a method has some
advantages in affording a direct pull, but the cost of construction would perhaps be larger than for a structure on
the banks. The illustrations show plans of supporting the
motor over the canal by cables which would be a practical
form of construction.

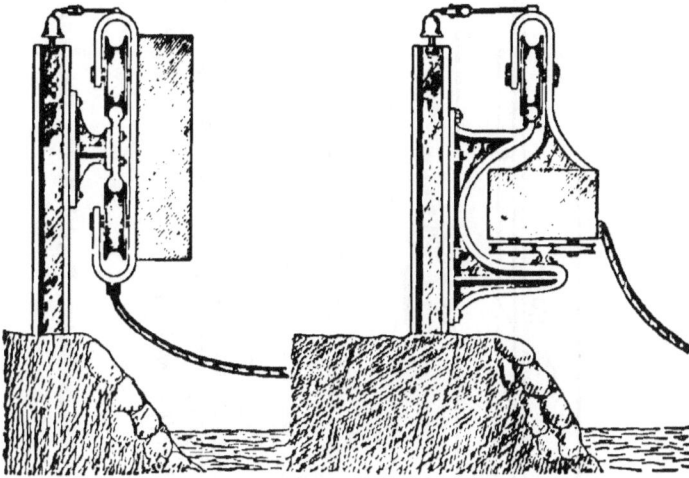

FIGS. 74 AND 74a.

In this method of canal boat propulsion, the light hauling
motors run upon a rail, suspended by suitable wire cables,
over the canal. Figs. 75 and 75a show transverse and
longitudinal sections of a canal equipped with this method.
Spanning the canal at intervals of from 50 to 75 feet are
heavy wire cables supported from poles erected upon the
banks. These poles are similar to the ordinary trolley
poles used for electric railways, but of stronger construction and stand about 8 to 10 feet above ground. Suspended
longitudinally along the route of the canal are two cables

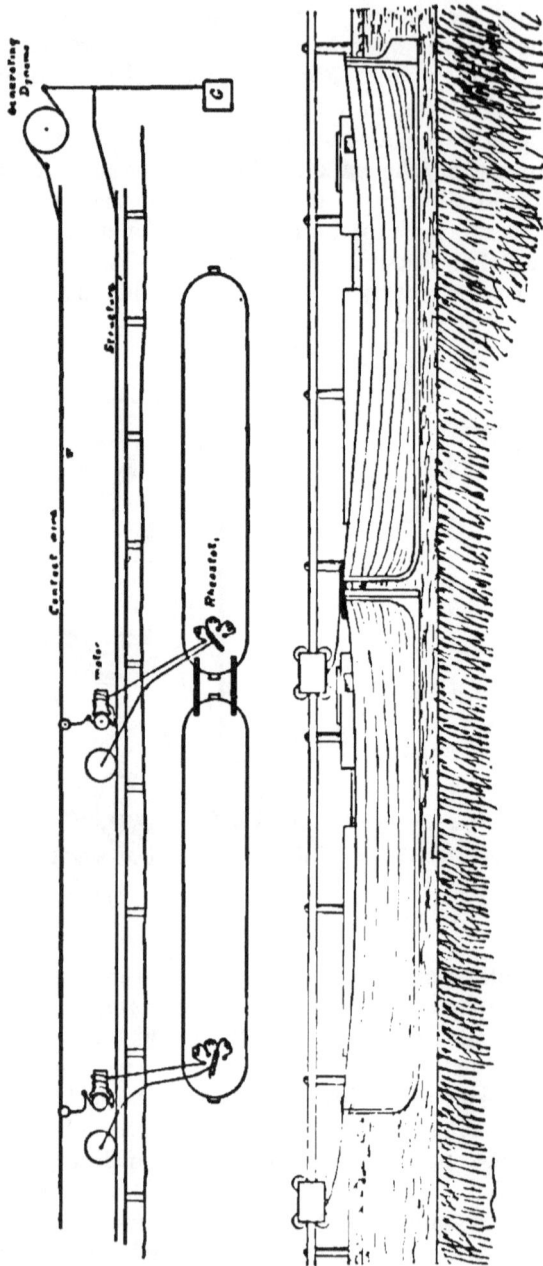

Fig. 74b.—General Arrangement, Sachs' Motor Haulage System

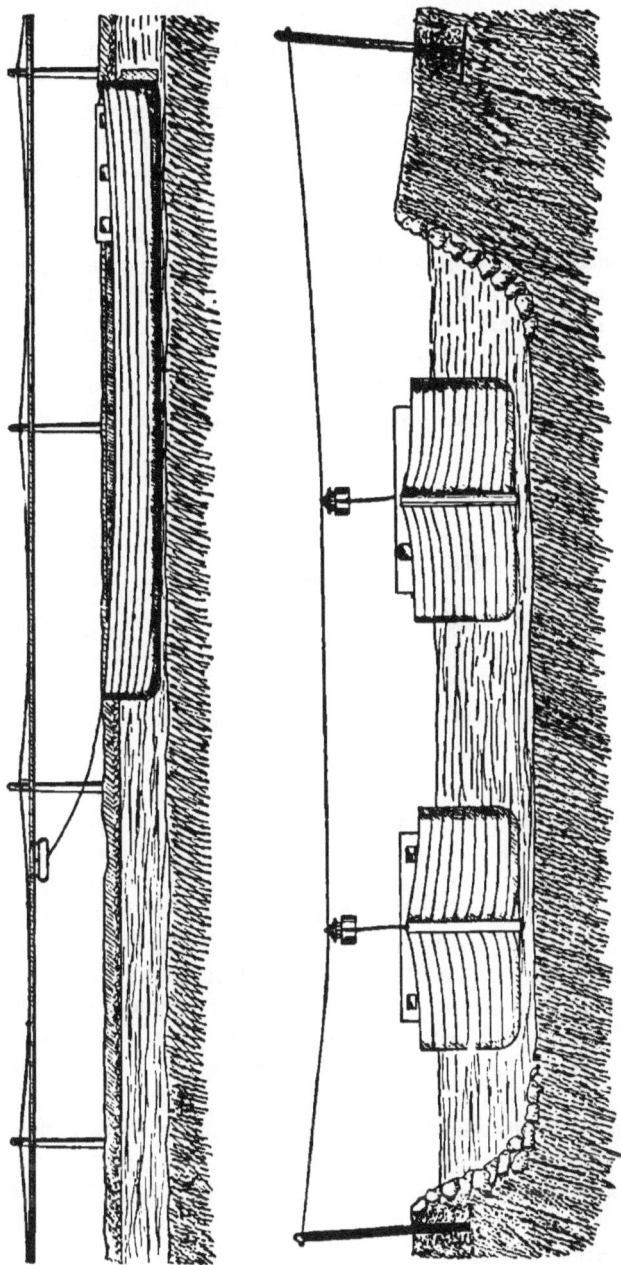

FIGS. 75 AND 75a.—SACHS' SYSTEM OF CABLE SUSPENSION ELECTRIC MOTOR HAULAGE.

which are fastened to, and supported by, the transverse
spans. The rails upon which the motors are run are supported by each of these longitudinal cables.

To understand better how the motor is supported, Figs.
75b, 75c, 75d and 75e are shown. Fig. 75b gives an end
view of a hauling motor supported upon the suspended
rail. A shows the transverse cable which supports the
longitudinal cable, B. To this longitudinal cable the rail, C,
is fastened at intervals of from 5 to 10 feet as shown. This
rail weighs about 50 pounds per yard. The contact wire is
supported beneath the rail and the circuit is composed of
the wire and rail. The motor carriage, D, is supported by
means of the rollers, E, which run upon the flange of the
rail, C, on each side of the supporting web. The carriage is

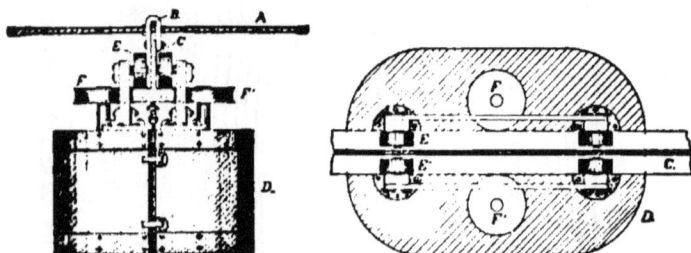

FIGS. 75b. AND 75c.

propelled by the flanged rollers, F F', which firmly grasp the
sides of the rail and are rotated by the motors on the carriage. More or less traction power can be obtained by
altering the pressure of the flanged rollers, F F', against the
rail. Instead of using an arrangement as shown, a rack
and pinion may be used for the smooth roller and rail.

Fig. 75c gives a plan view of the hauler carriage with the
web of the rail in section. Fig. 75d shows a side view of
the hauler carriage, rail and supporting cable. Fig. 75e
shows the motors and gearing within the carriage. The
gearing is so arranged that either or both motors can be
used. Various other methods of transmitting power from
the motor to the propelling rollers can, however, be used,
and it would, perhaps, simplify matters to use but one
motor instead of two.

As will readily be seen, this method of motor haulage has various advantages. Such an arrangement would not in any way interfere with the present operation of the canal and could be economically constructed and maintained. It will be noted that the propeller method can also be operated without any addition to the structure shown. The trolley

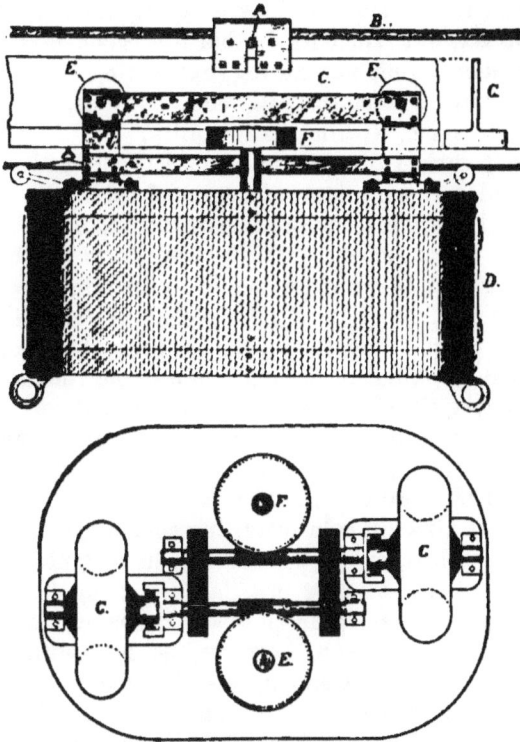

FIGS. 75d AND 75e.

carriage would be so arranged as to run upon the rail, and contact would also be made with the wire. Current would be taken to the motor on the boat by means of a flexible cable.

73. An arrangement shown in Fig. 76 would perhaps be possible on some narrow, continuously level canals. As will

be seen, a structure is built over the canal on which a high speed electric railroad can be operated, while the hauling motors for the boats in the canal are operated below the tracks as indicated.

Instead of using the arches over the canal, as shown, the cable method of suspension could be used and would be particularly applicable here.

Such an arrangement would cost a large amount of money to construct, but when its double use is considered and it is remembered that no grading, excavating or other work, required with a surface road, need be done, such a plan appears, to some, within practicability.

A great variety of structures and supporting rails could be devised, but some form of overhead cable suspension or single rail structure on the banks would, no doubt, be a desirable arrangement. With any form of structure used, it would be most advisable to operate the boats as shown in Fig. 74b. It will be seen here, that there are two boats coupled together as in the present mule method. Each boat is pulled by a single motor and the motors can either be regulated separately on each boat, or from one controller on the forward boat. Such an arrangement permits of erecting a light structure just strong enough to withstand the strain of one boat.

The chief difficulties in installing such a system of haulage, as described, would be met with in switching, so as to permit the boats to pass one another going in the same direction and in passing under obstructions, such as bridges. These requirements can, however, be met with by suitable switching devices and by erecting the motor supporting structure in such manner as to permit of its being attached to the bridge. The passing of boats going in the same direction, does not, however, seem to be an absolute necessity as the boats would be continually in motion at a constant speed. Although it would, perhaps, be an advisable arrangement on the Erie, it would not be so necessary on private canals. Suitable switching stations could be placed along the route so as to enable boats to pass one another going in the same direction.

Such a system of motor haulers possesses a great many advantages. It does not in any way necessitate any addi-

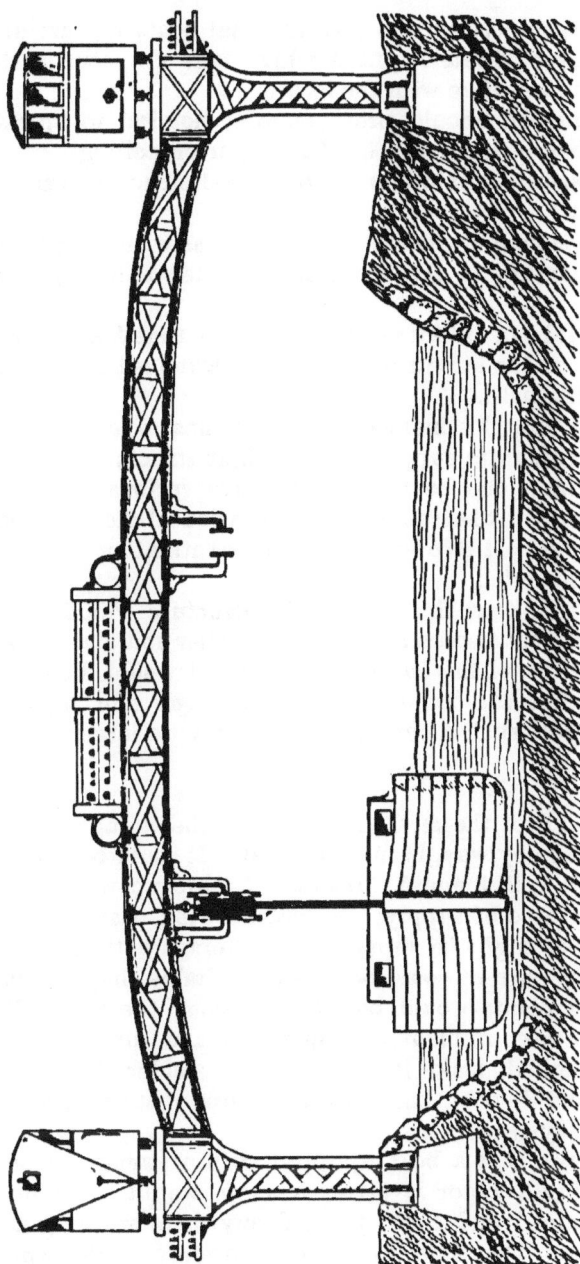

FIG. 70.—COMBINATION OF HIGH SPEED ELECTRIC RAILWAY AND MOTOR HAULAGE SYSTEM ON ONE STRUCTURE.

tional machinery on the present canal boats nor are any extra towing boats required which take up room in the canal and interfere with the locking.

The method should prove more economical to operate than other systems, whether hauling or propelling.

Any speed permissible by the canal structure can be obtained.

There is no disturbance of the water as with the propeller, and injury to the banks is prevented to a very great extent.

It would not interfere with the present methods of propulsion, as the structure could be so arranged as to clear the towing path.

As a whole, this method of hauling canal boats by small motors which are supported on a light structure and controlled from the boat, seems to have various points of superiority. It is true, it also has various difficulties, but these should all be overcome by proper attention to design and construction.

It is true that the cost of the structure would be in excess of that necessary for the propeller and also of that of one or two of the hauling systems. It would, however, appear that the various other advantages and superior efficiency of such a system would warrant a larger cost of plant.

74. In all of the methods described, the storage battery has not, in any way, been considered. It could be applied to classes 1, 2, 3 and 5. Instead of taking power from a central source of supply, the barge, towing boat or motor locomotive might be supplied with storage batteries from which the current could be taken. Suitable charging and changing plants would be situated along the line. The immense cost of plant and operation, and difficulties in such a method, make it necessary to consider the central station and distributing wires as a source of supply for any practical system.

· It is true that a boat equipped with a storage battery and having a motor revolving a screw, would be entirely self-contained and could proceed anywhere without any assistance. The same thing could, however, be done by a steamer, either barge or tug. Therefore, the only possible advantage would be one of cost, and in that respect the superiority would appear to remain in favor of the steamer.

Of the various methods of applying electricity to canal boat propulsion, the struggle for superiority seems to lie between some form of propeller and one of the various hauling methods. From the requirements necessary for a practical system, it would appear that a separate motor tug would be the best form of applying the propeller.

Of the various hauling methods, the operation of a small hauling motor on an elevated structure adjacent to the canal, towing or hauling the boat in the water and regulated from the boat, seems to be the most practical and feasible. Some form of cable suspension over the canal or single rail on each bank, would be, perhaps, the most practical form of structure to use; although the duplex structure possesses many points in its favor.

Aside from the question of economy, the exterior motor method certainly seems to have various advantages over the propeller, as will readily be seen by carefully considering both. When, however, the *cost of operation* of each of the two methods mentioned is also considered, the hauler method again seems to be superior to the propeller. In the following chapters an approximate estimate of the cost of installation and operation of both systems will be made.

CHAPTER XIII.

GENERATING PLANT AND DISTRIBUTION.

75. Before comparing further the methods of propellers versus motor haulage, let us look into the questions of generating and distributing power for canal boat propulsion. The practicability of any electric system will greatly depend upon the cost of operating the prime mover, and the efficiency of transforming the energy into electricity and distributing it to the points of consumption, where it is re-transformed into mechanical energy.

It is believed by many that the working pressure or voltage should not exceed 500 volts, and that preferably a simple two-wire system (either alternating or direct current) should be used. It may also be found that a direct current will prove most practical for the working circuit unless some efficient and simple single phase alternating motor should offer. The general electrical arrangement of the working or feeding circuit and the motors would be the same as in ordinary street railroad practice.

We can, however, use various systems of distribution to secure this working pressure, which will greatly be governed by the size of the canal, its location, kind of power available, etc. Whatever method of generation and distribution is used, it must permit of continuous operation 24 hours of the day.

76. One of the simplest forms of generation and distribution would be to use the methods employed in ordinary electric railway systems, by generating and distributing the electric current directly. As the distributing distance would be limited by the cost of copper, it would be necessary to erect numerous stations along the route of canals that are of any length. The generating machinery could be operated by either steam or water power, as found most practical.

77. Upon long canals, where it would be necessary to erect numerous stations, other methods of distribution could, however, be employed. Current could be generated in one or more large central stations situated at some cheap source of power, distributed at high tension, and reduced to the working pressure of the motors by transformers along the route of the canal. This plan permits of various modifications. We could, for instance, distribute direct current at high pressure and reduce to the working voltage by motor transformers along the route, or a single or multiphase alternating current could be used and transformed to a direct current of lower pressure at the rotary transformers.

In districts where cheap power and labor can be obtained along the route of the canal, the direct method of generation and distribution would, no doubt, be used. Where, however, cheap power could not be obtained along the entire route some modification of the high tension method would be adopted.

78. As the Erie and other canals in the higher latitudes are closed during part of the year, the most economical operation of plant could not be obtained if power were only furnished for canal boat propulsion. It would, therefore, be most practical and advisable to arrange the distributing and generating plant so as to operate at all times ; power being furnished for other purposes as well as canal boats. In the case of the Erie and some other canals, an electric railway could be operated parallel to the canal or over it ; and light, heat and power could be furnished by the same plant used to operate the boats. If an electric system is used it should also be applied to the operation of lock gates, etc., along the canal.

79. In comparing the two methods of electrical canal boat propulsion, we shall roughly assume, for the sake of simplicity, that we are distributing power by the direct method with stations located along the route of the canal about 20 miles apart (distributing 10 miles each way from the station), and using steam as a motive power. This will necessitate about 18 stations along the route of the Erie

canal. The station machinery, engines and dynamos, will be arranged in units of say, two or three hundred horse-power each and one-third of the generating plant will be surplus over that actually necessary to furnish power for boat propulsion ; which will be necessary as the system must be kept constantly in operation. We shall only consider the use of the plant for canal boat propulsion.

The cost of a combined steam and electric generating plant may be roughly placed at $100 per horse-power and coal to be consumed at the rate of 2¼ pounds per horse-power hour at dynamo terminals ; 20 per cent. will be allowed for loss in distribution. It next becomes proper to determine how far the expense will now be governed by the adoption of either the propeller or the motor haulage method, but before passing on to deal with that question, a comparison should be made between the steam driven electric genera-tors, just referred to, and the supply of power from Niagara to the canal. Messrs. E. J. Houston and A. E. Kennelly, electrical experts of high standing, have recently con-tributed to *The Electrical Engineer*, of New York, the subjoined data on the subject of Niagara power as com-pared with the highest class of compound condensing engines.

80. One of the most important questions of the day is, How far is it commercially possible to transmit water power electrically? It is recognized that the limiting commercial distance depends upon two associated inter-dependent factors ; viz., cost and electric pressure.

(1.) *Cost*, including the purchase and maintenance of the necessary machinery and wires together with the annual interest chargeable upon such expend'ture.

(2.) *Electric Pressure.* The pressure or voltage at which the line transmitting the power can be operated with con-tinued security to life, and assurance of permanence of supply, and permanent protection to the lines of conductor from lightning, weather and all disturbances.

It is clear that if reliable machinery could be purchased cheaply enough, and the conductors could be safely oper-ated at sufficiently high pressure, the Falls of Niagara could to-day stop steam engines in New Orleans, La., by

underselling their power.   There must be a certain radius
from Niagara, within which electrical power will carry
death and displacement to the ordinary steam engine, and
it is desirable to ascertain how far this radius may be
likely to extend.

We shall first assume that a steady transmission of
power is to be provided for, from Niagara Falls to various
cities along the banks of the Erie Canal, as shown in the
accompanying table.

Distance.	Miles.	K. W.	H. P.
To Buffalo...	15*	23,500	30,160
To Syracuse	164	7,500	10,060
To Schenectady.................	300	7,500	10,000
To Albany	330	15,000	20,110
To various points along canal, for barge propulsion		15,000	20,110
		67,500	90,500

* To outskirts of Buffalo city.

We shall assume that the engineering difficulties can all
be overcome by bare, overhead, tri-phase wires, at 35,350
volts receiving pressure, with step-up and step-down
transformers at each end of the lines.   It remains now to
consider only the question of cost.

TURBINES AND HYDRAULIC WORKS.

We require to estimate the annual cost of producing one
kilowatt steadily at the turbine shafts at Niagara Falls.
It has been stated that the estimated cost of the hydraulic
works at Niagara for a total of 119,000 H. P.,

Was.....	$10.50 per H. P.	or	$14.070 per K. W.	
If we add as much as 20 per cent. for any possible miscalculation...................	2.10 "		"	2.814 "
And also add cost of turbines at	5.00 "		"	6.703 "
	$17.60		$23.587	

Allowing 5 per cent. for interest on capital, 2½ per cent.
for superintendence and repairs, and 2½ per cent. for
depreciation, the total annual cost is 10 per cent. on investment, or $1.76 per H. P., or $2.359 per kilowatt, assuming
that the turbines are always running at full load.   This, however, is more than can reasonably be expected.   The average load under favorable conditions can hardly exceed 60
per cent. of the maximum load handled, making the annual

cost $2.932 per H. P. or $3.931 per kilowatt of average annual delivery.

Another line of reasoning is as follows :—It has been stated by Prof. Forbes that he did not continue to add cost to his generators, when $60 of extra expenditure in them did not increase their electrical output by 1 H. P.   Prof. Forbes also mentions that he took 5 per cent. as interest on investment, so that it seems clear that his estimate of the annual cost of producing one horse-power at generator terminals is 5 per cent. of $60, i. e., $3 per H. P. or $4.021 per K. W.; and eliminating the cost of generators, this would be about $3.72 per K. W. at turbine shafts.   However, as our purpose is to be conservative, it may be safer, in the absence of actual assurance, to take $4.00 per K. W. as the cost of power of turbine shafts per annum ($2.984 per H. P.).

GENERATORS.

The cost of alternators, as is well known, varies greatly with their size and capacity.   In sizes below 3 K. W. their purchasing cost may be $100 per K. W., while in sizes of, say, 50 K. W. they may cost $45 per K. W.   Finally in very large sizes, say of 4,000 K. W., their purchasing cost would probably be reduced to $8.516 per K. W. ($6.352 per H. P.), including exciters, and all station apparatus.

MOTORS.

The cost of motors may be expected to average slightly higher than that of generators, for the reason that they will frequently require to be made in smaller sizes (say, for 1,000 K. W.) than the generators.   The purchasing cost per kilowatt may be taken as $9.581 per K. W. ($7.146 per H. P.) with exciters and apparatus included.

TRANSFORMERS.

The step-up and step-down transformers would probably not cost more than $5.157 per K. W. ($3.847 per H. P.). Prof. Forbes has stated that offers as low as $3.52 per H. P. ($4.72 per K. W.) have been made to deliver such transformers.   These transformers may be conveniently regarded as forming part of the generators or motors at each end of the lines.   In this regard, allowing 0.96 as the efficiency of generators or motors, the purchasing cost of generators and all accessories becomes $13.888 per K. W., or $10.352 per H. P. Purchasing cost of motors $14.953 per K. W. or $11.154 per H. P.

The efficiency of complete generator and motor plants becomes 0.94 at full load, allowing 0.98 as efficiency of transformers.

CONDUCTORS.

It is obvious that the right amount of capital to expend in line conductors and construction, after the voltage and frequency have been decided upon, is that such additional expenditure just ceases to save its value in interest, by reason of the consequent reduced loss of power in the line. In cases where the expenditure in the line and construction is in proportion to the weight of copper in the line, this economical point must be attained when the annual charges of interest, repairs and depreciation on the whole line, are equal to the money value of the power lost in transmission along the line, which is Kelvin's law, and is applicable to alternating as well as to continuous currents and whether such currents have lag or not. Usually, however, the expenditure upon the line does not increase directly with the weight of conductor, and the first statement is, therefore, more generally applicable than Kelvin's law. It can be shown that with the prices above mentioned, the most economical weight of copper to employ on each main tri-phase conductor, will be about 20½ pounds per ampere per mile.

Allowing for the best quality of oil insulators, and bare wires, not exceeding No. 000 A. W. G., a first-class ordinary pole line can be erected for $1,000 per mile of line; and if this allowance be made for the line to each city, the combination will admit of a substantial and permanent structure in wood or in iron, capable of holding all the wires safely, and in a manner practically safe from lightning, yet readily accessible to linemen for repairs. By properly grading the distances between conductors, the antagonistic effects in the lines of capacity and inductance can usually be rendered negligible, so that the fall of pressure in the lines becomes that simply due to ohmic resistance.

By systematically transposing the wires of all circuits, say at every mile, their mutual inductance would be negligibly small and their capability of disturbing one another would be consequently annulled.

SUPERINTENDENCE AND OPERATING EXPENSES.

The following staff might be expected to handle the system.

One general superintendent and assistants................	$30,000 per annum.
One chief engineer...	20,000 "
Twelve assistant engineers...................................	36,000 "
Forty dynamo assistants......................................	40,000 "
Forty linemen and assistants	40,000 "
	$166,000
Office expenses, salaries, taxes, etc........................	34,000 "
Total...	$200,000 ".

Or $3.00 per H. P. of total capacity, or $2.681 per K. W.

ALBANY.

DISTANCE, 330 MILES; MAXIMUM DELIVERY, 15,000 K. W.

For a delivery with full load at motor shafts of............	1.000 K. W.
The delivery at line terminals.................................	1.064 "
And with a power factor of 0.9, the received current on each conductor is...	0.01931 amp.
Copper per mile on each conductor	0.3958 lb.
Total copper in line, per K. W. delivered..............	391.9 lbs.*
Approximate loss of energy in line...........................	0.862 K. W.
Approximate energy delivered at Niagara terminals.....	1.926 "
Approximate energy delivered at Niagara turbines........	2.048 "
Cost of 1.926 K. W. generator at Niagara at $13.888 per K. W........................	$26.750 per K. W.
Cost of 391.9 lbs. of copper in lines......................	48.987 " "
One kilowatt motor at Albany at $14.953...................	14.953 " "
	$90.690 " "
Cost of 330 miles of line at $1,000 for 15,000 K. W	22.000 " "
Total investment per K. W. delivered........	$112.600 " "
Annual cost of interest, depreciation and repairs at 10 per cent..................................	11.269 " "
Cost of 2.048 K. W. at turbine shafts at $4.................	8.192 " "
	$19.461 " "
Labor, superintendence and general expense...............	2.681 " "
Annual cost of delivery of one K. W. at sustained full load.	$22.142 " "
At average of 60 per cent. of full load, $27.53 per H. P. or	$36.903 " "

---

* The number of wires in the line to Albany would be nine, about No. 000 B. & S. gauge, three on each main conductor.

They next give corresponding calculations for Syracuse and Buffalo, and then summarize as follows :

SUMMARY.

Cost of delivering a max. of 22,500 k. w. at Buffalo...at $10.694	$240,610
Cost of delivering a max. of 7,500 k. w. at Syracuse..at 17.223	129,170
Schenectady.................. 7,500 k. w.................at 21.260	150,450
At Albany.......... ...... 15,000 k. w... ....at 22.143	332,130
Different points.............. 15,000 k. w.................at 17.830	207,450
67,000	$1,128,810

Average cost of delivery, full load. .....$12.475 per H. P. $16.723 per k. w.
Average cost of delivery 0.6 full load...$20.792 per H. P. $27.872 per k. w.
By Emery's tables the cost of generating
  steam power per annum with coal at
$3 per ton is for 308 *days of ten hours*.$25.27 per H. P. and $33.88 per k.w.
365 days of 20 hours...... ....$44.43 per H. P. and $59.56 per k.w.

This is with large triple expansion compound engines.

The foregoing results indicate that on the basis of prices and voltages assumed and detailed, the power of Niagara Falls can be transmitted to a radius of 200 miles cheaper than it can be produced at any point within that range by steam engines of the most economical type with coal at $3 per ton. That Niagara power can maintain at Albany a large day and night output cheaper than steam engines at Albany can develop it, but that for power taken at Albany for 10 hours per diem the best steam engines have somewhat the advantage over Niagara unless exceptionally favorable conditions of load could be secured for Niagara power.

These conclusions are, of course, entirely dependent upon the reliability of the prices, voltages and estimates as detailed above. The prices, however, appear to be conservative, and are probably in excess of the latest market values. This would also appear from the statement recently published that the Cataract Construction Company has contracted for the privilege of right of way for conducting lines along the Erie Canal banks, by supplying power for barges at the annual rate of $20 per H. P. ($26.81 per kilowatt) whereas the above figures make the average cost of production $20.792 per H. P. ($27.872 per k. w.) on the basis of an average output of 0.6 full load ; so it would appear that the Niagara Power Company are in the possession of better prices. Moreover, if the voltage delivery on the long lines, here taken as 35,350 at the receiving end, can ever be safely increased, a marked reduction in the cost of delivery would naturally result.

The broad conclusion to which an inquiry of this nature inevitably leads, is that while under ordinary conditions the commercial limit of electrical transmission of power from water powers of less than 500 kilowatts can hardly exceed fifty miles, the radius at which it will be profitable with good fortune and management, to electrically transmit a water-power aggregating 50,000 K. W., or more, is, perhaps, to-day, two hundred miles, and that it might be commercially advantageous for such a large water power to undersell large steam powers at twice this distance with no profit, in order to reduce the general expense upon delivery nearer home. The reason for this difference in the transmission radius between small and large water powers, lies obviously in the fact that electrical and hydraulic machines can be built and purchased much more economically in large sizes than in small, so that the cost of producing and of maintaining one kilowatt is very much less for large than for small water powers.

We may add here that these interesting calculations have been variously discussed and criticised, and that several dissenting opinions of weight have been expressed. Messrs. Houston & Kennelly have nevertheless adhered to the figures presented, maintaining their accuracy, and our readers will probably soon have the opportunity of seeing in how far this confidence is justified.

CHAPTER XIV.

RESISTANCE OF CANAL BOATS.—COMPARISON OF COST,
PROPELLER *vs.* HAULER.

81. In order to compare the two methods of propelling
canal boats—the propeller and hauler methods—we shall
consider the installation of an electric system of boat propulsion on the Erie Canal, assuming 350 miles as the total
length. We shall assume that the canal is open for 215
days of the year and that we have at least one foot more in
depth of water throughout the canal, giving a standard
depth of 8 feet, and that the locks are all lengthened so that
two boats can pass through at one lockage. (See page 101.)

Upon this improved canal we shall assume the operation
of 600 barges (300 going in each direction), each carrying
240 tons of freight. For the sake of simplicity, we shall
assume that these boats are continuously moving at an
average speed of 3 miles per hour. Such would not actually be the case, as a number of boats would be waiting at
the various locks and, therefore, taking no power. The
percentage of all the boats on the canal that would be
locking at one time would be small. The total amount of
energy required to operate each boat the entire length of
the canal, would not be seriously affected by the locking;
but to allow for the lockages and also for the smaller
number of boats at the beginning and ending of the season,
we shall assume that all the boats are taking power for 200
days. The full number of boats would gradually be
reached at the opening, and decrease at the closing, of the
canal. It will, therefore, be necessary to consider the
operation of the power plant for the full time of 215 days.

Resistance of Canal Boats.

82. As we are going to operate the ordinary barges at
present in use (98 feet long, 17½ feet wide and 6 feet

draught, loaded to 240 tons) by electric methods, let us
look into the resistance met with by these boats on the
Erie Canal under different conditions. The law of boat
propulsion which could be applied to boats in a broad
expanse of water is not applicable here, and the contracted
waterway enters greatly into the computation. Ex-State
Engineer Elnathan Sweet has given in a paper read before
the American Society of Civil Engineers, March, 1890, a
table compiled from various experiments made by him on
the Erie Canal. This table is shown below.

TRACTION EXPERIMENTS ON ERIE CANAL, 1878, NEAR CANAJO-
HARIE, WITH CANAL BOAT HENRY L. PURDY.

Velocity.		Sections.		Ratio of Boat to Prism Section.	Depth of Water.	Immersion of Boat.	Weight of Boat, lbs.	Weight of Cargo, lbs.	Resistance.				Submerged Surface of Boat.
									Observed.		Computed.		
Feet per sec.	Miles per hour.	Of Boat Immer., Sq. ft.	Of Prism. Sq. ft.						Lbs.	H. P.	Lbs.	H. P.	
1.86	1.27	105	450	4.26	7 ft.	6 ft.	130,000	467,600	297	1.	278	.94	
2.71	1.85	"	"	"	"	"	"	"	575	2.83	590	2.92	5970 Sq. ft.
2.86	1.95	"	"	"	"	"	"	"	634	3.29	657	3.41	
2.04	1.89	105	525	5.	8 ft.	6 ft.	130,000	467,600	292	1.08	280	1.08	
2.90	1.50	"	"	"	"	"	"	"	343	1.37	326	1.3	
2.30	1.57	"	"	"	"	"	"	"	381	1.60	356	1.48	5970 Sq. ft.
2.88	1.96	"	"	"	"	"	"	"	594	3.11	658	2.92	
3.06	2.09	"	"	"	"	"	"	"	645	3.58	649	3.50	
2.14	1.46	113.75	525	4.61	9 ft.	6.5 ft.	130,000	521.120	362	1.43	344	1.83	
2.94	2.00	"	"	"	"	"	"	"	632	3.37	656	3.68	5950
3.14	2.14	"	"	"	"	"	"	"	734	4.19	783	4.19	
1.95	1.33	122.5	525	4.2	9 ft.	7 ft.	130,000	574.740	350	1.24	330	1.17	
2.94	2.00	"	"	"	"	"	"	"	678	3.62	718	3.83	5960
2.65	1.81	"	"	"	"	"	"	"	602	2.90	607	2.92	

These experiments were made in the fall of 1878, with
the boat "Henry L. Purdy," which fairly represents the
average canal boat on the Erie Canal. The boat was towed
by horses and the resistances were measured by a dynamo-
meter and are shown in column marked "observed." The
resistances shown in column marked "computed" were

deduced from the equation $R = \dfrac{0.10303 \ s \ v^2}{r - 0.597}$ units being

in feet, pounds and seconds. The slight differences between the observed and computed resistances were caused
by the rapidly varying power exerted by the animals.

The boats similar to the one experimented with have a horizontal cross-section of a little over 1,600 square feet above the 6 foot immersion line, and carry 25 tons for every half foot submersion beyond 6 feet. The submerged surface of these boats when drawing 6 feet is 2,870 square feet; when drawing 6½ feet, 2,980 square feet; and 3,090 square feet when drawing 7 feet of water. The resistances at various speeds for boats of similar model can readily be computed from the data obtained from the table and by using the formula (also given in the paper) of Scott Russell and Du Buat for the resistance of boats in narrow channels of water, viz.: $R = \dfrac{A \, v^2 \, s}{r + B}$ in which

$R =$ Resistance of boat.

$s \ =$ The submerged surface of boat.

$v \ =$ The velocity of boat.

$r \ =$ The ratio of the channel's water section to immersed section of boat.

A and B constants determined by experiment.

Mr. Sweet found the most probable values of A and B, for the model of boat used, to be,

$$A = 0.10303. \quad B = -0.597.$$

The equation therefore reads $R = \dfrac{0.10303 \ s \ v^2}{r - 0.597}$ as was used in the table; the units being in feet, pounds and seconds.

For velocities in miles per hour, this becomes,

$$R = \frac{0.2215 \, s \, v^2}{r - 0.597}.$$

Using the latter equation, the table on page 172 has been compiled showing the resistance of the ordinary canal boat, as used by Mr. Sweet, at one mile per hour with various boat immersion and canal prism sections and depths of water.

RESISTANCE OF CANAL BOATS AT ONE MILE PER HOUR.
COMPUTED BY FORMULA

$$R = \frac{0.2215 \; s \; v^2}{r - 0.597}$$

Speed	Immer. Section of Boat	Depth of Canal Water.	Sections.		Ratio of Boat to Prism.	Immer. Surface of Boat	Resistance.	
			Boat.	Canal Prism.			Lbs.	H. P.
1 mile per hour. 1.46 ft. per sec.	6 ft.	7 ft.	105.	450	4.28	2870	173	
	6 ft.	8 ft.	105.	525	5.	2870	145	
	6½ ft.	8 ft.	113.75	525	4.61	2980	164	
	7 ft.	8 ft.	122.5	525	4.28	3090	183	

It will be noticed that the resistance is about 15 per cent. less in 8 feet depth of water than in 7 feet.

Taking the resistances shown in this table at one mile per hour, we can readily find the resistance for any other speed for the same conditions. As the resistance of boats through water increases as the square of the speed, $R = L\,v^2$ when R = the resistance, L = pounds resistance at one mile, v = velocity in miles per hour. As the power necessary to overcome this resistance increases as the cube of the speed, $H.\,P. = P\,v^3$ when $H.\,P.$ = horse-power to overcome resistance of boat, P = horse-power necessary at one mile per hour, and v = velocity in miles per hour.

We can now very readily get the resistance of the ordinary canal boat as used on the Erie under the conditions previously named; viz.: 8 feet depth of water, 6 feet immersion of boat, speed 3 miles per hour. By taking the resistance at one mile per hour under these conditions at

145 pounds or $\dfrac{212}{550}$ horse-power, as shown in the table, we

find that at 3 miles per hour, the boat would encounter a resistance of 1,305 pounds, or it would take 10.4 horse-power to move the boat. Let us say, in round figures, 10 horse-power.

83. It has been found most economical, in operating boats on the Erie Canal, to connect them in pairs, as already described. We shall, therefore, endeavor to adhere to this general plan and arrange the electric methods so as to operate the boats in pairs and, also, so as to facilitate locking.

For the propeller method we shall consider the plan shown in Fig. 54, in which a small tug is used, pushing one freight barge and pulling two, making a fleet of three barges each carrying 240 to 250 tons, with a total of 750 tons.

For the hauler method we shall consider the plan as outlined in Fig. 72, in which a small motor is operated on a light structure adjacent to the canal, each motor pulling a single boat and being regulated from the boat. It will be more advisable to operate only two boats by this plan.

Propeller Method.

84. As we are going to operate 600 boats (300 each way), we will have 200 three-boat fleets or trains. Allowing 10* horse-power as the power necessary to overcome the resistance of each boat, and because the resistance increases with the surface, we will have 30 horse-power as the total power necessary to overcome the resistance of each fleet of three boats (excluding the tug).

There will, however, be power used above this which is necessary to propel the tug, and that which is lost in the slip and friction of the screw. As is well known, the loss of power in a screw propeller, in ordinary cases, amounts to from 15 to 30 per cent. of the power applied at the shaft. With the steam canal boats used on the Erie Canal at present, this loss is much larger, amounting to from 50 to 80 per cent. (with an average of about 65 per cent.). By sub-

---

* The allowance of 10 H. P. per boat necessary to propel it at three miles per hour is low. As has been shown previously, this is the power necessary to overcome the actual resistance of the boat at the above speed and does not take into consideration the additional resistance caused by the propeller. The augmented resistance due to the screw propeller would, no doubt, add greatly to the actual hauling resistance which is considered here. The propeller increases the resistance of the forward boat by preventing the water from fully reacting upon its stern, and causes an increased resistance in the rear boats by the stream that is thrown astern by it.

stituting an electric motor for the steam engine on the canal boat, this loss could perhaps be somewhat reduced. The ordinary slippage of a screw propeller increases when other boats are towed, and it is doubtful whether the losses in slip and friction of the propeller screw could be reduced to any great extent in an electrical method if the screw and motor are located on the canal boat.

If, however, the propeller screw is placed upon a small tug, as in this case, and revolved at a higher speed, thereby allowing the use of a smaller screw which is kept constantly immersed, this loss could be somewhat reduced. It will perhaps be fair to assume that in the case under consideration, the combined losses in the screw and power to propel the tug, will amount to 45 per cent. of the power at the motor shaft. We must, therefore, have 45 per cent. more power at the motor shaft than is finally necessary, in order to overcome the resistance of the canal barges; and as this is 30 horse-power, the power necessary at the motor shaft on the tug would be 55 horse-power. Allowing a loss of about 10 per cent. in the motor, we have, in round numbers, 60 electrical horse-power necessary at the motor terminals. This would be the power taken by the motor when the boats are loaded to a depth of 6 feet immersion, each carrying 240 tons.

Upon the Erie, the boats going east are generally fully loaded, while those going west carry but a small amount of freight. We may fairly assume that the average power necessary at the motor terminals for westward fleets will amount to about 20 horse-power. Although this may not be strictly correct, it will answer in the present calculation.

As there are 100 fleets going in each direction we would require a total of 8,000 horse-power delivered at motor terminals. Allowing 20 per cent. loss in distribution, we find that it would be necessary to have 10,000 horse-power at the dynamo terminals at the generating stations in order to propel these 600 boats. This will give an efficiency of about 40 per cent. for the system from the dynamo to the actual power used to propel the boats. If we have 18 stations along the canal, each station would have an output of about 555 horse-power. The machinery installed, however, would be about 50 per cent. in excess of this, or about

800 horse-power. The entire generating capacity would be about 15,000 horse-power.

85. Taking the coal consumption at 2½ pounds per electric horse-power hour delivered at the dynamo terminals, we get a total coal consumption of 25,000 pounds per hour. At this rate we get about 3.5 per horse-power hour at the motor shaft and about 6½ pounds per effective horse-power hour.

As a rough approximation of a propeller plant under the above conditions, with a steam generating plant, we may, therefore, take the following :

Cost of Electric Propeller Plant.

Distributing system (poles, wires, feeders, etc.), 350 miles at $6,000 per mile, - -	$2,100,000
Generating plant (buildings, steam and electric), 15,000 H. P. at $100 per H. P.,	1,500,000
Propelling tugs, 200 at $4,000, - - -	800,000
Superintendence, engineering and incidentals, - - - - - - -	200,000
Total cost, - - - - -	$4,600,000

We can also roughly approximate the cost of operating this plant for 215 days.

Cost of Operation.

Labor to operate tugs, one man per tug, -	$75,000
Labor (station, line, superintendence and office), - - - - - - -	100,000
Coal, 60,000 tons at $3 per ton, - - -	180,000
Interest on investment, 4 per cent., - -	184,000
Maintenance, depreciation, etc., } oil, water, waste, insurance, etc., } - -	225,000
	$764,000

Let us call this $765,000.

At this approximate figure, power will cost about 2 cents per horse-power hour delivered at the motor shaft on the tug. This will make the cost of propelling a fleet of three boats, going at three miles per hour, each loaded to 240 tons by the method described, as follows :

 36.6 cents per mile,

 12.2 "   "   boat mile,

 $128.10 for entire trip, Buffalo to Albany (loaded).

Going in the opposite direction (lightly loaded) propulsion will cost :

 12 cents per mile,

 4 "   "   boat mile,

 $42.00 for the entire trip west from Albany to Buffalo.

This would make an entire round trip under these conditions cost about $170.

The cost of propulsion for each of these boats for an entire round trip would be about $57.

Hauler Method.

86. Let us now compare with the above the method of hauling as outlined in Fig. 72. In this method we shall operate but two boats in each fleet ; each boat pulled by a single motor. As we have 600 boats, we will have 300 two-boat fleets or trains, 150 going in each direction. As each boat encounters a resistance of about 1,305 pounds (10 horse-power at three miles per hour) the motors must exert an effective pull to overcome this. Allowing a loss of 20 per cent. in the gearing of the hauler, we find that each motor must furnish 12½ horse-power at the shaft. Each of the motors would, however, be of about 15 horse-power capacity. Taking 10 per cent. loss in motor, as before, we get 14 horse-power of electrical energy necessary at the terminals of each motor.

This would be for loaded boats. For light boats going in the opposite direction, we shall assume, as in the previous case, about one-third of this or about 5 horse-power.

As we have 300 boats going in each direction, and as each boat is operated by a single motor, it will be necessary to furnish a total of 5,700 horse-power at the motor terminals. Allowing 20 per cent. loss in distribution, the power needed at the terminals of the generating dynamos will be 7,125 horse-power. The entire capacity of the generating plant would be about 10,700 horse-power. The efficiency of the entire system from dynamo terminals to power actually used to propel the boats is about 58 per cent.

Although we could operate this plant with a smaller number of stations than in the previous case, we shall assume that the same number of stations are to be used, but of smaller capacity. Each station would, therefore, have an output of about 400 horse-power and a capacity of about 600 horse-power.

At the rate of 2½ pounds of coal per horse-power hour delivered at the dynamo terminals we also get the same as before, 3½ pounds per horse-power hour at motor shaft, but the coal consumption per effective horse-power hour would be less than in the previous case; amounting to 4½ pounds per horse-power hour.

We may consider the following as a crude approximation of the cost of the hauler method. The supporting structure considered is a light, single support on each bank of the canal, or else some form of cable suspension as shown in Fig. 76.

Cost of Electric Hauler Method.

Structure and distributing system, 350 miles, at $10,000 per mile,	$3,500,000
Generating plant, 10,700 horse-power at $100 per horse-power,	1,070,000
Motor haulers, 600 at $1,000 each,	600,000
Superintendence, engineering, etc.,	200,000
Total cost,	$5,370,000

The cost of operating this plant under the same conditions as before would approximate about as follows :

Cost of Operation.

Labor (station, line, superintendence and office),	$100,000
Coal, 42,750 tons at $3.00 per ton,	128,250
Interest on investment at 4 per cent.,	214,800
Maintenance, depreciation, etc., oil, waste, insurance, etc.,	200,000
	$643,050

In round figures, $645,000.

Taking these figures we may form an estimate of the cost of propulsion by the hauler method. The cost per horse-power hour delivered at motor shaft will be about 2.6 cents.

It will, therefore, cost 10.8 cents per boat per mile (21.6 cents for each pair of boats or fleet). This will make the cost of the entire trip east of two boats, each loaded to 240 tons, going at 3 miles per hour, about $75.50. Going in the opposite direction the cost of propulsion would be about one-third, or about $25.00. The entire round trip of two boats by this method would cost about $100.00; or a continuous round trip of a single boat would cost $50.00.

Comparison : Propeller vs. Hauler.

87. From these crude estimates we have, therefore, the following :

	Propeller.	Haulage.
Cost per boat mile (east), - -	12.2 cents.	10.8 cents.
" " " trip (east), - -	$43.00	$38.00
" " " " (east and west),	57.00	50.00

or as a comparison of the total operating expense of 600 boats for 200 days at 3 miles per hour, we have,

Propeller, $765,000 ;   Hauler, $645,000 ;

making a difference of $120,000, or 15 per cent. in favor of the hauler.

We have considered in both cases a very crude arrangement of generating and distributing plant, and in the latter method particularly, the general plan would be somewhat uneconomical. In fact if the plant were properly planned a larger saving could be shown in favor of the hauler method. It must also be remembered that we have considered that the resistance of the boats is the same in both cases. This is not a fact, as the resistance of the boats operated by the propeller tug would be greater than by the hauler method. If we take these facts into consideration, and furthermore, if such plant were installed by the State and no profit was expected on the invested capital, it will readily be seen that the advantage in favor of the hauler method would be still greater.

Comparison With Present Methods.

88. With this rather rough approximation it would hardly be fair to make any depreciatory comparison with the present methods. With the actual plant that would be installed with either method, the cost of propulsion would no doubt be brought down to quite some extent. It must also be remembered that but a very small and limited output has been considered. An increase in the number of boats propelled and a continuous operation of the generating plant—power being also used for other purposes— would also greatly decrease the cost of propulsion. An approximation of the cost of propulsion by the present methods would, however, be interesting.

At the present speed of 2½ miles per hour in 7 feet of water, we may take the following as the cost of propulsion per boat mile by the present steamer-and-consort method. Each barge carries 240 tons and steamer 180 tons.

Coal, 25 pounds at $4.25 ton, - - -	5	cents.
Engineer's pay and board, - - -	2	"
Oil, waste, etc., - - - - - -	1.5	"
Interest, repairs, insurance, etc., on propel- ling machinery, - - - -	1.5	"
Total cost per boat mile, - -	- 10 cents.	

We cannot, however, compare this but must reduce it to the same conditions as in the electric method, namely 8 feet of water and 3 miles per hour.

Referring to the table on page 172, we find that there is a saving of about 15 per cent. in the resistance of a boat loaded to 6 feet when moving in 8 feet of water instead of in 7 feet. Taking the coal consumption in 7 feet at 25 pounds per boat mile, at 2½ miles per hour, we find that in 8 feet it would be 21¼ pounds. As the power necessary to propel a boat increases as the cube of the speed, at three miles per hour, in 8 feet, we would get about 37 pounds per boat mile. As the other conditions will alter but slightly, we will assume them to be the same; and we now have

Coal, 37 pounds at $4.25 per ton,	- -	8 cents.
Engineer's pay, board, etc.,	- - -	2 "
Oil, waste, etc., ' - - - -	- -	1.5 "
Interest, repairs, insurance, etc.,	- -	1.5 "
Total cost per boat mile,	- -	13 cents.

This figure is, however, more likely to be about 15 cents per boat mile. As the cost of towing by horse or mule is higher than by steam it will be unnecessary to consider it in such a comparison.

Taking then the various methods, we have the following as an approximate estimate of the cost of propelling the ordinary canal boat on the Erie Canal at three miles in 8 feet of water, loaded to 240 tons.

Steam propeller,	- -	13	cents per boat mile.
Electric "	- - -	12	" "
" hauler,	- -	10.5	" "

It would appear from the above, and from the fact that the estimates are very crude and that the cost of operation can be decreased to a large extent by an economical and correct design of plant, that an electric system can be economically operated. It would also appear that the hauler method had the advantage in all directions. It is difficult to make any positive statement as to ultimate cost of operating such a method, but the cost could perhaps be decreased 20 per cent. below the figures given even if we limit ourselves simply to numerous generating stations using coal, and not to the cheap power from Niagara, which, if the calculations of Messrs. Houston and Kennelly be correct, can be delivered even in Albany on terms of more than an equality with the energy from the best triple expansion engines on the spot.

CHAPTER XV.

Propulsion: Resistance of Boats and Propellers;
Paddlewheels and Screws.

89. The subjects treated in this chapter relate not only
to electric launches, but to the other classes of boats dealt
with in the present volume, including canal boats. It has
been deemed proper, to discuss them broadly, leaving
fuller treatment to books on the specific topics that are
here passed in review. As will have been noted, some
data on the subject, as applied to canal boats, has been
given in the previous chapter.

Boats can be propelled through water in two ways; first,
by exerting a direct pull or strain on the boat sufficient to
overcome its resistance; second, by the use of a propelling
instrument on the boat which acts upon the surrounding
water in an opposite direction to that in which the boat is
to be moved.

The first plan has been illustrated in the previous chapters by the cable towing and motor hauling methods on
canals. Such methods of propulsion are limited to canals
and rivers. The second plan of propulsion is, perhaps, the
more important and, in fact, the only way of giving motion
to a boat through water, excepting as above. The principle
involved in any marine propeller is the projection of a
quantity of water in a direction opposite to that in which
the boat or vessel is to move, and of such mass as is necessary to overcome the resistance of the boat.

The resistance of a body, such as a boat, moving through
a broad expanse of water depends upon quite a number of
conditions which are quite variable. The total resistance
encountered by a boat is made up of the following:

First.—Frictional resistance due to the surface friction
between the water and the skin or surface of the boat.

Second.—Wave making resistance due to the cutting of

the water by the bow of the boat which sets up waves called waves of displacement.

The frictional resistance is by far the most important in boats of ordinary good lines and moderate speed. It depends upon the area of immersed surface and its degree of roughness, and varies about as the square of the speed. It is also affected to some extent by the length of the boat, but is not affected by its form or proportions except in extreme or unusual cases. Even with high speeds, the frictional or skin resistance is the largest part of the total resistance encountered by the boat.

The wave making resistance depends upon various elements, but particularly on the form and proportions of the boat. With a boat of good model and fairly fine lines, at fair speed, the resistance caused from waves would amount to but a very small percentage of the total resistance.

There is, however, a limiting speed for a boat of any particular model, above which the resistance increases very rapidly. In some large poorly built boats a slight additional resistance is caused by eddies formed at the stern. In boats of good lines, however, this is quite inconsiderable.

Various methods have been used to deduce the resistance of a boat of any particular model or form at a certain speed, all of which are based for their accuracy upon certain practical data obtained from experience with other boats of similar model. It is quite difficult to lay down a general formula of resistance for all models of boats, as the wave resistance of a boat of narrow beam going at a certain speed would be less than that of a boat of the same displacement and midship section going at the same speed, but of broad beam. It will, therefore, be seen that although the immersed surface of the two boats may be about the same, their resistance at a certain speed may be different.

It must also be understood that wind, tide and waves in rough water, increase the resistance of the boat. The resistance from these causes cannot be approximated and can only be arrived at from actual results and experience under the various conditions.

The resistance of ordinary boats of good model at

moderate speed, in fairly smooth water, varies as the square of the speed ; and the power necessary to overcome this resistance, as the cube of the speed. This would apply to boats without reference to the mode of propulsion, except in the case of certain propellers, such as the screw, which have a tendency to augment the resistance under certain conditions.

The resistance of a boat, however, also depends upon whether the boat is traveling through a broad expanse of water or along a confined waterway—such as a narrow river or canal. The resistance encountered by a boat, in a confined and narrow waterway, is very much greater than that experienced by the same boat in an unlimited expanse of water and depends upon the relation-between the section of the boat and that of the waterway through which it is passing. It will readily be understood that if the channel section is small as compared with that of the boat the resistance encountered by the boat will be greatly increased even at very low speed. At comparatively high speeds the resistance caused by the contracted waterway is experienced to a very much greater extent.

90. We shall not endeavor to go into all the details of boat resistance under various conditions. It may, however, be well to give an empirical rule used to approximate indicated horse-power required for screw propelled steam boats or vessels. It has been found from practice that an allowance of about 5 horse-power per 100 square feet of wetted surface at a speed of 10 knots per hour will approximately give the indicated horse-power necessary at the engine cylinder. This allowance can, however, only be made with boats of fair model and good lines, and at moderate speed in free waterways.

The above has been used in connection with larger boats than would generally be found among the types considered here. For boats of small size the allowance of horse-power for the given speed and surface would no doubt be slightly larger. The proper allowance can no doubt be easily deduced from experiment.

With some alteration this rule could be applied to the electrical energy necessary under similar conditions.

We may assume that the losses from the engine to the screw amount to 25 per cent. of the total indicated power. With an electric motor the losses will probably be half of this, or, let us say, 15 per cent. We may, therefore, deduce the electrical energy necessary at the motor terminals as follows :

$$\left(\frac{S}{10}\right)^{3} \times \frac{4\,A}{100} = \text{horse-power electrical energy at motor.}$$

Where S = Speed in knots per hour.

A = Area wetted surface of boat in square feet and allowing 4 horse-power per 100 feet of immersed wetted surface of boat at 10 knots per hour.

In order to reduce this approximation of the energy necessary to propel a certain boat, to the resistance met with by the boat (leaving out the losses in the engine and screw), we may assume that the loss between the indicated horse-power at the driving engine and the real effective propulsive power used to overcome the resistance of the boat is from 50 to 60 per cent. This would, therefore, give us a rate of about $2\frac{1}{4}$ horse-power per 100 feet of wetted surface at a speed of 10 knots per hour. Reducing this to pounds we get approximately a resistance of about 80 pounds per 100 feet of wetted surface at 10 knots per hour. Therefore, to get the resistance in pounds approximately :

$$\left(\frac{S}{10}\right)^{3} \times \frac{80\,A}{100} = R \text{ (resistance in pounds) ;}$$

$$\text{or, } \frac{R \times F}{33000} = \text{effective horse-power.}$$

S = Speed in knots per hour.

A = Area in feet of wetted surface.

F = Speed in feet per minute.

The allowance made for each 100 feet of immersed surface of the boat is not at all constant, but would vary with the surface and model of boat. With the ordinary model, such as used in electric launches and boats, it would form a rough approximation of the power required. We are assuming here, however, that the boat is propelled by means of a screw. With boats that are hauled, as in systems of canal boat towing, this would not apply. In such cases the absolute resistance of the boat in passing through the surrounding water must be deduced in a manner similar to that in a previous chapter, and the horse-power necessary to overcome the resistance will vary as the cube of the speed.

91. Under all ordinary conditions, excepting as above, boats are propelled by means of a propelling instrument or mechanism located on the boat.   The propeller may be either a paddle, a screw or a jet, but the action of the propelling instrument upon the water is the same in all cases.   It will readily be seen that an amount of work must be done in an opposite direction to the direction of the boat in order to overcome the resistance of the boat and cause it to move forward.   As has been previously stated, all propellers work on the principle of projecting a certain mass of water, equivalent to the resistance of the boat, in an opposite direction to its motion.

The velocity of the projected stream is always greater than the actual velocity imparted to the boat.   As the water is a yielding medium it will readily be understood that there must always be a certain slip of the projected stream.   Therefore if V be the velocity of the stream projected by the propelling instrument, and v be the velocity of the boat, then $V - v =$ velocity of the stream with respect to the water.   This is called the slip of the propeller.   The stream thrown from the stern of the boat is the slip.   If the entire speed of the boat were equal to that of the projected column of water there would be no stream thrown astern and no slip.

Although the jet, the paddle or screw propellers could be utilized for the propulsion of electric boats, the screw is particularly adapted to the propulsion of such boats on account of the high speeds obtainable with the electric motor.   It has also been shown by practice that the screw has various advantages for such craft.   We shall particularly deal with the screw propeller, but it will be well to also give a general description of the two other forms of marine propellers.

92. The jet propeller is, perhaps, the simplest in principle and operation, but it is more or less impractical and very

inefficient. It operates by throwing sternward from the boat a stream of water which has been drawn in at the bow or sides and passes through pipes in the boat. Between the inlet and outlet a pump, turbine or other device is located for imparting a velocity to the water, which, being in an opposite direction to the direction of the boat, causes a certain thrust which must be equivalent to the resistance of the boat.

93. The paddle in its simplest form, consists of a number of flat boards or paddles, radially secured to a circular frame, revolved at its centre, so that, as the paddle wheel is revolved, the paddles push the water back in an opposite direction to that in which the ship is moved. If the water were unyielding, its action would be analogous to that of a rack and pinion, and the distance traveled would be $\pi \times$ the pitch circle of the paddles or circle of centres of pressure. As the water yields, however, the action is the reverse to that of an undershot water wheel.

This form of propeller was the first to be used for mechanical boat propulsion. It is confined now chiefly to boats for inland navigation. There is another form of paddle wheel which, although acting in the same manner, is much more efficient. This wheel is called the feathering paddle and is so arranged that the floats, instead of being bolted as in an ordinary paddle, are pivoted on an axis parallel to the axis of the wheel. The paddles are connected by rods with an eccentric which is so set, with regard to the axis of the revolving wheel, as to cause the paddle or float to be nearly vertical when entering the water. By this means the floats always have a direct sternward action on the water and not an oblique action as is the case with the radial paddle.

The feathering paddle is, perhaps, the most efficient marine propeller and is used to-day to a great extent for inland and shallow water navigation. It has, however, various objections, particularly that of bulkiness; and for rough water it is entirely impractical.

94. The screw is the most generally used marine propeller and is the only practical and efficient propelling

instrument for the great majority of the boats described
in this book. The action of the screw is more complex
than either of the other forms of boat propellers.

The general principle of the action of the screw propeller can be compared to that of a bolt and nut—the
water represents the nut. If the nut be fixed and the bolt
be revolved, a forward or backward motion will be imparted to the bolt at right angles to the direction of
rotation. If a blade of metal be wound endwise upon a
cylinder or shaft so as to form a screw thread upon the
shaft and revolved in water surrounding it, the same effect
will be produced; that is, the water will take the place of
the nut and a motion parallel to the axis of the screw will
ensue. All screw propellers are based upon this principle.

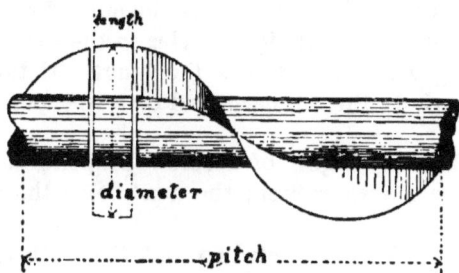

FIG. 77.

It is not necessary, however, that the blade should make
a complete revolution around the shaft. If a slice be cut
from such a thread, as shown in Fig. 77, a piece of blade
will remain. If two such threads are wound around the
shaft, two blades will remain, and so on. Such a slice
with two or more blades forms a screw propeller. Each
blade represents a thread of the screw. Two or more slices
of threads or blades are used in order to decrease the
surface friction between the surface of the blade and
the water.

The pitch of a screw propeller is the distance measured
along the axis of one complete revolution of the screw
blade. The pitch of ordinary screws varies from one to
two and one-half times the diameter.

The disc area is the circle swept by the edges of the blades as they revolve.

The thrust of a screw propeller is the power exerted in pounds on a line parallel to the axis.

The developed area or blade surface is the area of all the blades added together.

Fig. 78.

The projected blade surface is the sum of all the blades measured on a plane at right angles to the axis.

The length of the screw is the length of the slice cut from the whole thread, measured on a line parallel to the axis.

The "leading edge" of a screw propeller is the edge that strikes the water first; the next being the "following edge."

When the blade is twisted so that the pitch varies along the blade, the screw is said to have a varying pitch. If the pitch increases toward the periphery of the blade, the propeller is said to have an expanding pitch. If it is larger at the hub, the pitch is said to increase radially.

Fig. 79.

Several forms of propeller screws are shown in Figs. 78, 79, 79a and 79b.

If the screw propeller were worked in a solid medium, as in the case of the nut and bolt, the velocity of the boat driven by screw would equal P (the pitch) $\times$ R (the revolutions) of the screw. But, as will readily be understood, the boat does not generally advance at this rate. With the screw, as with other marine propellers, there is always a certain slip of the projected column of water which is lost in a stream thrown astern by the propeller.

FIG. 79a.

Let P be the pitch of screw and R number of revolutions, then V (speed of propeller) $= P \times R$.

If now the speed of the boat be v, then

$$V - v = \text{slip of screw};$$

$$\text{or,} \quad \frac{V - v}{V} \times 100 = \text{per cent. of slip.}$$

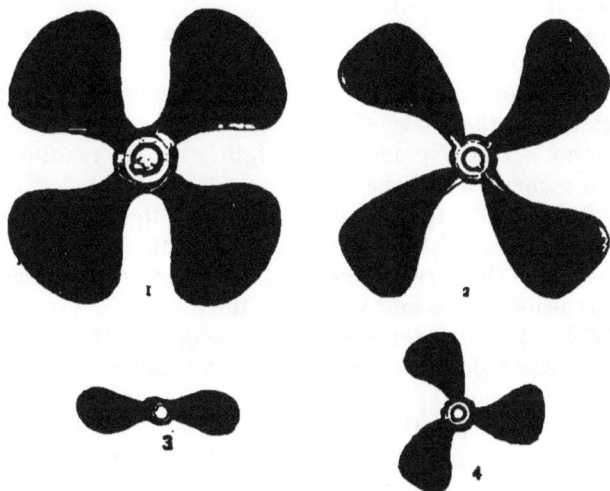

FIG. 79b.

This is, however, the apparent slip of the screw, that is, the slip relative to still or quiet water. As, however, the boat causes a current to follow in her wake, this current must be taken into consideration in getting at the real slip. In some cases it has been found that v was larger than V, caused by the following current being so strong as to entirely counteract the apparent slip. This is called negative slip.

Although it would appear that this would be a case of added efficiency, such is not the case. To create the following wake, a certain expenditure of energy has been necessitated which can only be partly returned. It is very seldom met with in well proportioned boats of the class here treated of. It must, however, be remembered that in calculating the real slip, the velocity of the following wake must be taken into consideration. The slip generally spoken of is the apparent slip.

The real slip represents the true value of the backward velocity impressed on the water by the propeller, and is not to be regarded as an evil characteristic but the contrary. Absence of real slip is a sign of inefficiency. It will readily be seen that the thrust transmitted to the boat by the screw shaft will vary with the velocity of the water thrown astern.

The laws of slip of screw propellers have been deduced from experiment as follows :

First.—For the same screw, slip increases with the resistance of the boat.

Second.—The slip increases slightly as the revolutions of the screw are increased.

Third.—The slip increases with the pitch. Therefore screws of fine pitch should be most efficient.

Fourth.—The screw should be made as large as possible with relation to the mid-section of the boat.

Fifth.—The slip decreases as the length of the screw or as the area of the blades with a fixed diameter increases.

Screws of smaller pitch have generally been found to have less slip than those of coarse pitch. As such screws generally revolve at a high speed, the loss from surface friction between screw blades and water becomes greater. Screws to work at high speeds, however, can have less area than those at low speeds. With electric launches and most other boats (herein described), high speed, low pitch, screws would be used.

In designing a screw it must, however, be borne in mind that in order to get the best results the screw must be kept immersed. If the propeller screw breaks through the surface of the water, it carries down with it a certain amount of air which is a cause of inefficiency. This is particularly the case with large screws. The number of blades does not seem to affect the working of the screw except as above stated. Three or four, however, give less vibration than two.

95. It is particularly imperative that the lines of the boat should be such as to permit a free flow of water to the screw. It has also been found that the action of the screw increases the ordinary resistance of the boat by preventing the water that has been displaced by the bow from reacting on the stern as it should. This augmented resistance amounts to but a small percentage in well designed boats. Energy is also wasted in throwing water radially from the propeller instead of backward. It is to prevent this centrifugal action that the propeller blades on some screws are inclined astern. Thin blades and small pitch are preferable for such boats as electric launches. It has been pointed out that the slip of a screw propeller decreases with the pitch, but it must also be kept in mind, that as the screw revolves, the friction between the blade surface and the water must increase with higher speeds with the same blade area. With large screws revolving at high speeds this surface friction has been found to amount to several per cent. of the total energy used by the screw in propelling.

The action of the screw is to drive sternward a column of water whose area is equivalent to the area of the screw disc minus the boss or hub. The velocity with which this area is projected by the propeller $= P \times R$. As has been shown, however, this cannot be the velocity imparted to the boat by the propeller; but the velocity imparted to the boat by the propeller will be the speed of the propeller less the slip. Let $v =$ velocity of boat in feet per second, $V =$ speed of propeller or velocity of projected stream per second, $A =$ area of the column of water projected in feet. Then $A \times V =$ volume in cubic feet of water projected per

second.   Taking the weight of a cubic foot of sea water at
64 pounds and gravity at 32 pounds, we get:

$$(T) \text{ Momentum of stream} = \frac{A \times V \times 64}{32}(V-v) =$$

$$2 A \times V (V-v).$$

This is the thrust of the propeller exerted along the
shaft of the screw in pounds, and varies as

$$A \times (P \times R)^2.$$

To move the boat this thrust must overcome the resistance of the boat, and the speed at which the boat will then
move will be as previously shown.   The thrust × the speed
of propeller = total work, and resistance of boat × speed
of boat = useful work.   The horse-power necessary at the
screw shaft outside of the boat varies as $A \times (P \times R)^2$ or

$$H. P. = \frac{T \times S}{33000},$$

S = speed of propeller per minute.

It will readily be understood that the velocity of the
following wake, if there be any, must always be taken into
consideration when the velocity of the projected stream is
to be deduced.

It will readily be seen, also, from the above formulæ
and general principles, how the diameter, pitch, thrust and
horse-power of a screw propeller can be approximated
for a certain resistance of boat.   In order to get the
energy necessary at the motor terminals, the losses in
transforming the electrical into mechanical energy and
transmitting it to the propeller must be considered.   If
a constant resistance be deduced for a certain model of
boat at a certain area and speed of propeller and for a
standard area of wetted surface, the propellers necessary
for other boats of the same model can be roughly deduced
from the above data.   As, however, screws of different
diameter and pitch have a varying percentage of slip, $V-v$
will vary.   As has been shown, the resistance of boats
varies with conditions, and although the screw propeller
has been investigated by many, we are still compelled to
resort to experimental tests with screws upon boats of
various model to get constants to be used in a computation.

The formula given, although quite rough and approximate, will, however, allow reasonable deductions to be made within the limits stated.

Those who are particularly interested and desire to further investigate the propulsion of boats and propellers, will find of use various papers read before the Institution of Naval Architects, particularly by Dr. Froude ; the book on "Marine Propellers," by Sydney M. Barnaby ; and various other books and papers on naval architecture and marine engineering, which treat of the subject fully. Among these also are the writings of Prof. Rankine, A. E. Seaton, W. H. White, Scott Russell and Du Buat.

CHAPTER XVI.

THE MISCELLANEOUS USES OF ELECTRICAL POWER FOR BOAT PROPULSION AND CANALS.

96. The electrical propulsion of boats from some central source of generation or distribution is not limited to the classes already discussed. Such methods can also be applied to boats on other waterways. In the illustration (Fig. 80) a feasible plan for an electric ferry is shown, in which a motor on the boat receives current from overhead wires suspended transversely over the river or other channel. The trolley or contact wire is suspended from a cable spanned across the river attached to the towers or poles on each bank. Where a railroad bridge, for example, crosses the river adjacent to the ferry, the wire can be supported directly beneath the bridge. These contact or trolley wires are arranged so as to permit a small trolley carriage to run thereon, and means are provided to keep the carriage securely upon the wire at all times. The boat would be connected with the carriage by a flexible cable which would permit sufficient lateral motion and would be connected with the motor revolving the propeller. The boat could be started, stopped and regulated in speed and direction by any suitable form of controlling or regulating device on the boat.

Such an electric ferry could, no doubt, be readily constructed across ordinary rivers and operated in conjunction with an electric railway. Its operation might be somewhat difficult across rivers having strong currents, but with sufficient slack cable this difficulty could be overcome. Although such an arrangement might possibly be operated beneath the Brooklyn Bridge, it is not intended to be operated on so large a scale ; but it certainly would seem to be a feasible arrangement for some small ferries that are located in the vicinity of electric railways from which current could be taken.

97. Another arrangement is shown in Fig. 81. In this plan current is supplied to the boat by a submerged electric cable which is connected with the source of electric supply on the shore. As the boat proceeds, the cable is let out, and taken in again as the boat returns. The reel for winding the cable is preferably located on the shore, as shown. In some cases, however, there would be no necessity for winding the cable. This form of electrical boat could, no doubt, also be practically operated under certain conditions.

At this point may be mentioned the plan patented by Mr. H. P. Wellman, of Catlettsburg, Ky., for the electromagnetic mooring of ferry boats, canal boats, etc. Broadly stated this plan consists in placing upon the boat and the wharf, alike, a series of mooring surfaces, as indicated in Figs. 82 and 83, one of the surfaces serving as a "keeper" to the other, so that the boat is held at her wharf upon approach and is released when ready for departure. The scheme does away with ropes, snubbing posts, etc., and should prevent bumping. It would answer

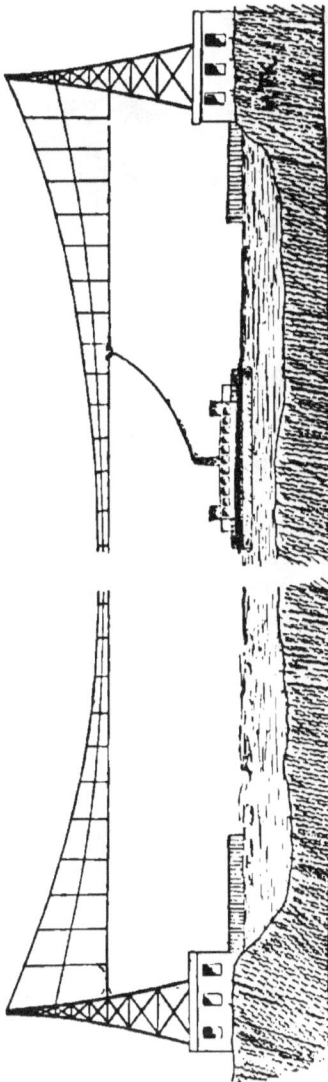

FIG. 80.—A METHOD OF ELECTRIC FERRY OPERATION.

best in still waters, and where there is little rise and fall
in tide.

Instead of holding the boat to its mooring, quay or dock
by suitable hawsers, as has been the practice, electro-

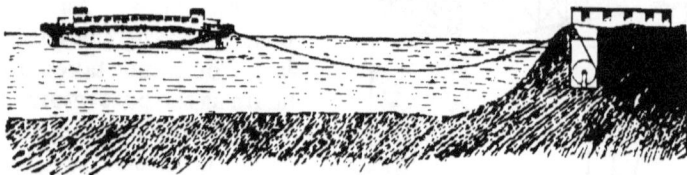

Fig. 81.—Electric Ferry Boat Operated by a Submerged
Cable.

magnetic means are used to accomplish this from the boat
itself. In one of the modified forms of this method which
is here shown (Fig. 82) electro-magnets are placed upon
the boat with their poles flush with the mooring side of
the boat which in the illustration is the bow. The wharf
or dock at which the boat lands is provided with a sheet
or apron of iron which is so placed as to abut and come
in contact with the poles of the magnets upon the boat
when a landing is to be made. Ordinarily the electro-
magnets upon the boat are not energized ; but when it is

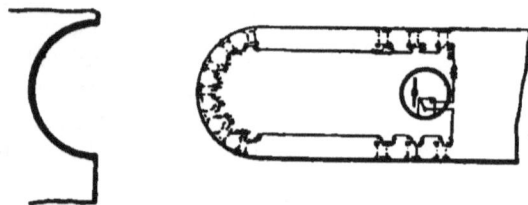

Fig. 82.

desired to moor the boat, the captain or pilot can at once
energize the magnets by connecting them with a source of
electricity, which may be the electric light dynamo on the
boat. As the magnets on the boat will abut the apron on
the dock, which will act as a keeper for the magnets, the
attraction between the magnets and keeper will be sufficient

to hold the boat. Instead of using magnets upon the boat only, they may also be placed upon the dock and supplied with suitable current, or the boat may be equipped with an iron keeper at its mooring portion and the magnets placed upon the dock. It would appear that this novel application of electricity would promise practical success particularly where frequent landings are made, as in ferries, etc.

98. Along a canal or river where an electric system of propulsion is used with distributing and contact wires paralleling the route of the waterway, the various bridges, lock gates, lifts, hoists and other machinery requiring

Fig. 83.

power should also be electrically operated. When canals pass through large cities, movable bridges are generally constructed across the canal. Such bridges, whether swing or lift, should and can all be operated electrically, and such apparatus is already in use in various parts of the world. Where a canal passes through a country of varying level, either locks or other devices must be provided to pass the boats from one level of the canal to another. Where locks are used, it is usual, in most cases, to use power to operate the lock gates. The electric motor would certainly be a most satisfactory substitute for any of the present methods. The boats could also be hauled into the locks by electrically operated hauling machinery. On some canals, such as the Morris or the Delaware and Hudson, the boats are trans- ferred from one level to another on cars or cradles which are moved by means of an inclined track by means of a chain. In Fig. 84 a cradle, car and boats are shown, as used on the Morris canal, New Jersey. The car is raised

FIG. 84.—INCLINED PLANE OF THE MORRIS CANAL, AT BLOOMFIELD, N. J.

and lowered by means of the chain to which it is attached. The application of electricity to this work is very simple and satisfactory. Direct lifts or elevators are also used upon some canals, which might just as easily be operated electrically as by steam or hydraulic power.

99. Upon the canal system extending from Lake Biwa, Japan, to the ancient city of Kyoto, the boats are hauled up and down an incline on cradle cars by means of a Sprague 50 horse-power motor driving the cable drum. This inclined railway connects the two canal sections, between which there is a fall of 120 feet, and upon which it was first intended to use locks. When the incline was adopted, it was intended, also, to use water-power direct, with a wheel at the foot, but in the present plan, the power-house is driven by Pelton wheels, and the Sprague motor is placed at the head of the incline. The total length of the inclined railroad is 1,800 feet with a grade of 1 in 15. The same power from the canal is being utilized for industries in the old city of Kyoto.[1]

100. A very large field of utility for the electric motor will also be found in the loading and unloading of boats and barges upon a canal operated electrically.

The electrical illumination of the towpaths, locks, etc., along a canal will also be a decided step in advance over the existing methods. The navigation of the Suez Canal at night by means of portable arc light plants has increased the facilities for traffic, and the income, of that canal enormously ; and where current is already used for power purposes the employment of a part of it at night is highly feasible. As an example of what can be done in this direction, it may be mentioned that the Baltic Canal will hereafter be illuminated by means of 25,000 incandescent lamps arranged along both banks. Each lock is also to have 12 arc lights, and electric lights are to be used for signaling purposes. The generating plant is to be placed in special power-houses at Holtenau and Brunsbuttel. This represents a plant of from 2,000 to 2,500 horse-power.

The Manchester, England, ship canal is electrically illuminated, and the proposed ship channel from the Gulf of Mexico to the city of Mobile is to be similarly lighted.

1. See "Electricity in the Far East." By W. Stuart Smith ; *Elec. World.* Vol. XXIII. No. 1, Jan. 6, 1894, p. 5 *et seq.*

In the latter case the channel is 30 miles long, and the alternating current has been adopted for the work.

101.   An electrical equipment has recently been furnished by the Canadian General Electric Company, of Toronto, to the Dominion Government, for the operation of the Sault Ste. Marie Canal locks.   It consists of a pair of "W. P. 50" motors for each pair of gates and each pair of valves, operated form either side of the canal.   In the case of the gates, the mechanical arrangement is the same as in a hydraulic equipment, with the exception that the hydraulic cylinder and piston are replaced by a pair of screws connected to the crosshead and operated by the motor.   In the case of the valves, which are of the horizontal butterfly type, the operation is performed by means of a crank on the end of the valve shaft, which is capable of being operated through an arc of 45 degrees, by means of a vertical screw driven by the motor.   The prime power for the generating plant is supplied by horizontal turbines, the water being taken from the upper level and discharged into the lower level in the usual manner.   As might be expected, a lighting plant is included in the system. Such work would soon become common with the general adoption of electricity for canal boat propulsion.

CHAPTER XVII.

STORAGE BATTERIES, MOTORS AND DYNAMOTORS.

102. Many references have been made in this volume to storage batteries, and as their use is important for the various purposes of electrical navigation, a few remarks upon some general characteristics and a few details as to special forms employed in this branch of work will not be out of place.

Reference has also been made to some of the differences between primary and secondary batteries, in favor of the latter, and the reader may consult again Chapters I. and II. It should be pointed out here, for the benefit of those who have not before made a study of electricity, that a primary battery is one which when exhausted can only be replenished, refreshed and renovated by a new supply of some of its constituents, such as the zinc or the solution in which the solid elements are immersed. These renewals are not only expensive—zinc, for example, being many times as costly as coal—but to effect them is generally a great nuisance and annoyance. To-day there are millions of such primary cells in use for telegraph, telephone, alarm, and other purposes; but the whole tendency of the times is to abolish them and to substitute either storage battery current or current generated by means of a small dynamo or motor-generator. From this obvious state of facts it may fairly be inferred that where heavy currents are necessary, to drive motors, for example, the primary cell is not likely to prove more desirable and efficient than in the departments where it is now less and less employed to furnish small currents for very light work. At the same time, it is not true that while primary batteries need frequent renewals, the storage battery is free from them, or needs only a charge of current to keep it in a condition to do its duty. Plates "pasted" or unpasted have been known to drop to pieces; buckling and sulphating are not unfamiliar idiosyncracies; and the acid solution evaporates. All these and other qualities or tendencies of decay are less noticeable now than they once were, and can be minimized by a little intelligent care; but to talk as though they were unknown, and as though storage batteries were free from the weaknesses of youth or the ailments of old age would be criminally to deceive the unwary.

At the present time storage batteries are in use successfully in hundreds of central stations and isolated lighting plants, where they are merely placed on shelves or floors and charged or discharged under favorable conditions of quiet and repose. Their use in street cars has been much less successful, owing, as a rule, to the great jolting or "washing" they undergo, the sudden strains due to abnormal calls for current and incessant starting or stopping, and also to the frequent handling in the transfer from power plant to car body and back again. In launch work we have to deal with conditions midway between those of station use and of car service. On a boat the batteries are retained under the seats or along the hull all the time and are charged in situ. They are hardly subjected to such heavy strains in starting the vehicle up from a dead rest, and when once the boat is started it is far less apt to make frequent stops, having usually a free course with definite landing places wide apart. Moreover, the motion of a boat through quiet water is far more tranquil and smooth than that of a car traversing even the best track that was ever laid. For these reasons, a storage cell might easily succeed afloat that had proved a failure under all the disintegrating influences experienced on land ; and for these reasons also, the storage battery has an enormous field of usefulness before it in launch propulsion. In canal boat propulsion also, the availability of the storage battery as an adjunct, either on the canal bank or upon the propeller, is not to be overlooked, and we must bear in mind the many uses to which current, cheaply generated, can be put along a canal, as evidenced, for example, by its use upon the Canal de Bourgogne already referred to.'

1. Chap. XI., page 125.

Modern practical storage battery use dates from the work in France of Planté, who used plain lead plates immersed in dilute sulphuric acid. These plates, however, as is well known, were very slowly "formed," so as to be fit for service; and it was not until Faure introduced the practice of applying the "active" material to the face of the plates, and thus prepared the battery for use by

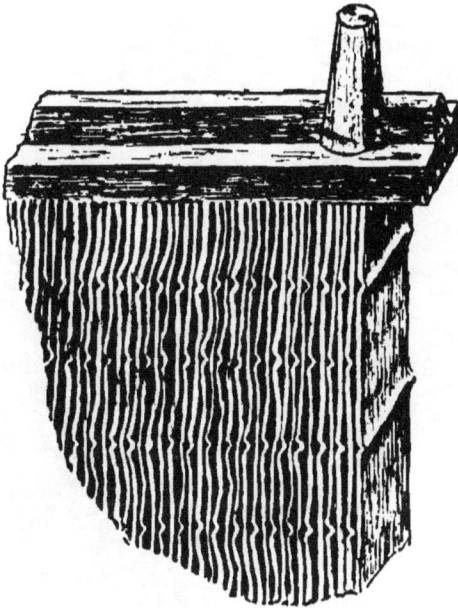

Fig. 85.

mechanical means instead of those strictly electrical, that the storage battery became really useful. The Planté and Faure types still remain the most prominent; but there are endless varieties and modifications of the lead battery form, as well as many others that are quite distinct and different. Only a few forms will be noted here, as examples, and for further information and data, our readers are referred to books devoted specially to the subject of storage batteries.

103. There are still not a few persons entitled to consideration who believe that the broad Planté method is the correct one, if properly and perfectly employed, upon which to construct storage batteries. The weak points in the original Planté type of cell are well known, and it was to overcome these and to increase the capacity as well as

Fig. 86.

the life of the cell that the Electric Power Storage Co., of New York, some time ago undertook a series of experiments carried out by Dr. Leonard Paget. In order to obtain a large surface for the action of the electrolyte a construction was adopted, as shown in Fig. 85, which represents a section of a plate with the upper connecting strip and connection plug; while Fig. 86 shows the cell complete and the manner of connecting the plates. As will be seen, the plates are built up of lead strips held in place at their upper and lower ends by horizontal bars which are cast over the ends of the lead strips, which are dovetailed for that purpose. The original form of the lead strips is a continuous ribbon, $\frac{1}{16}$ of an inch thick, 1 inch wide and weighing 1 pound per square foot. This is run through a machine which cuts it off to the required length and dovetails the ends. At the same time, the lead ribbon is nicked at intervals of $\frac{3}{4}$ of an inch, as shown in Fig. 85. When the strips are assembled to form a battery plate, these nicks, about $\frac{1}{16}$ inch in depth, form an offset which leaves a space between adjacent strips so as to leave free play for the circulation of the electrolyte.

The initial formation of the lead plates so made up is effected in a bath which converts the lead into lead oxide of lamellar structure, as distinguished from the usual granular form, with the result that it clings closely and intimately to the plate, so that no movement or buckling of the plate causes a dropping off of the active material. The plates are also subjected to a treatment by which it is claimed that the lead cores of the strips remaining after the formation are made impervious to action by the battery electrolyte.

The top of each plate into which the lead strips are cast is made of a composition not attacked by the acid and carries a dowel or taper stud which fits into a corresponding hole in the strip connecting the positive and negative plates of each cell. When the connecting strip is put in place a drop of solder is run into each connecting point and the joints thus made give additional conductivity. Each cell is connected to its neighbor by a connector bar which fits over a similar taper stud on the connecting strip.

320 AMPERE-HOUR CELL.

Discharge Rate.				100 per cent. capacity.			P. D.	
Unity	30 amps.	for 10 h.		100 per cent. capacity.			2.05 to 1.78	
1.46	46	"	" 6 "	92	"	"	"	"
1.07	62	"	" 4 "	83	"	"	"	"
3.25	105	"	" 2 "	70	"	"	"	"
5.08	106	"	" 1 "	53	"	"	"	"

6 ampere-hours per pound of Pb.

These cells are built in various sizes and capacities from 70 to 5,000 ampere hours, and the above table gives the results of a capacity test at different rates of discharge. At the normal rate the cell has a capacity of 6 ampere-hours per pound of lead. The 280 ampere-hour cell with connectors has a resistance of 0.00416 ohm. The internal resistance of the cell alone does not exceed 0.0001 ohm. No records are available of the use of this cell in navigation, but it would appear to be well adapted for the purpose.

104. The best known battery of the distinctive Faure type in use in America is that made by the Consolidated Electric Storage Company, of New York and Philadelphia, whose battery was used with such eminent success in the electric launch work at the World's Fair.[1] The plates in this battery are of the familiar perforated grid form, the positive plate being filled with red lead which is turned into peroxide by the current; and the negative plate with litharge, which is reduced to spongy metallic lead. In this form some marked improvements have been seen of late years. We give herewith an illustration of the cell (Fig. 87), and the subjoined table will show the capacity of the cell in various sizes :—

TABLE OF AMPERE-HOUR CAPACITIES AT VARIOUS RATES OF DISCHARGE, AND TIME IN HOURS.

Type.	No. of couples.	Rate of discharge.	Ampere hours.	Time of discharge in hours.	Rate of discharge.	Ampere hours.	Time of discharge in hours.	Rate of discharge.	Ampere hours.	Time of discharge in hours.	Rate of discharge.	Ampere hours.	Time of discharge in hours.
5 S..	2	1.	50	50	8.	40	13.	4	33	9.	5.	32	6.5
11 S..	5	2.5	125	50	7.5	100	13.	10	87	9.	12 5	80	6.5
17 S..	8	4.	200	50	12.	160	13.	16	140	9.	20.	130	6.5
15 L...	7	7.	400	57	14.	325	23.	25	310	12 5	35.	300	6.5
C. S..	1	2.4	24	10	8.	22	7.5	4	90	5.	6.	18	8.

105. We come next to what is known as the "Chloride" accumulator, with which, already, considerable launch work has been done in Europe, and which has been

1. For full description of this see Chapter III., pages 35–42.

similarly tried in this country, at Buffalo, and elsewhere. The makers are the Electric Storage Battery Company, of Philadelphia. It is shown in Figs. 88 and 89. It derives its name from the fact that the plates are made up of

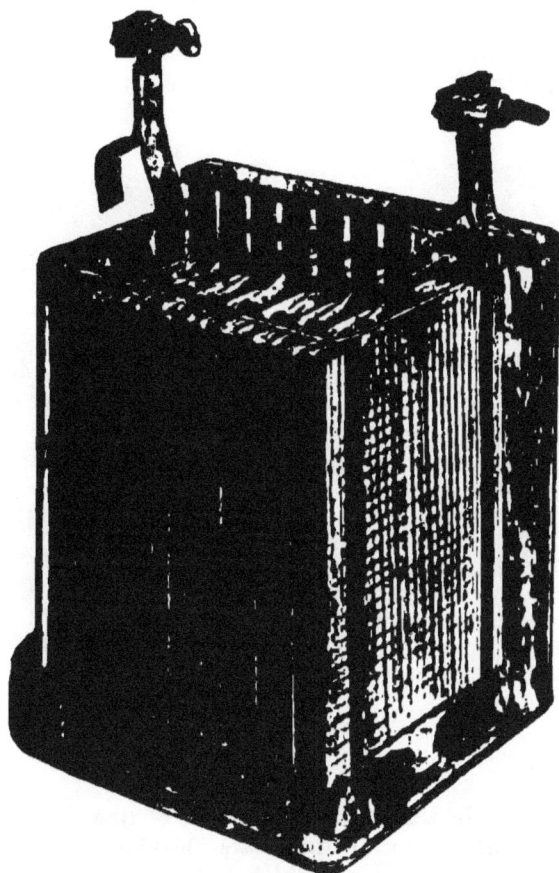

FIG. 87.

tablets cast from fused chloride of lead and zinc, which are held rigidly by a frame of antimonious lead. When so cast, however, they are not ready to be used, as the material, in this condition, is unfit to become active in a secondary battery. To make the plates active a chemical change is effected in the chloride tablets by means of a bath of chloride of zinc, in which the plates are immersed in connection with a slab of metallic zinc. The arrangement forms, in fact, a primary battery, the zinc acting as a positive and the tablets as a negative element. The electro-chemical action which results draws the chloride of zinc from the tablets by simple solution in the bath and also withdraws the chlorine from the chloride of lead and

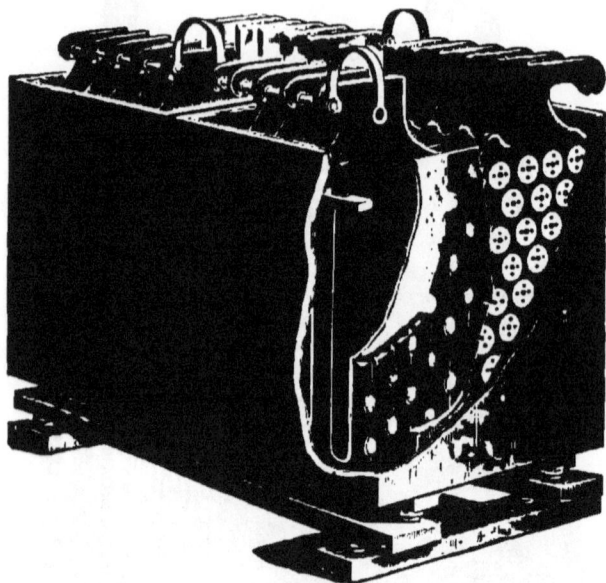

Fig. 88.

fixes it with the zinc, forming chloride of zinc. The latter is then washed out of the plate, leaving the mass of crystallized metallic lead, which is immediately available as active material in a storage battery.

Our engraving (Fig. 88) shows the Chloride cell as now constructed. As will be seen, it consists of a negative plate with round tablets of active material, which are perforated in order to permit of the free circulation of the battery fluid. The negative is separated from the positive plate, first by a separator, made of wood, soaked in insulating compound, and perforated to correspond with the location of the tablets in the plate. The perforations, it will be noted, are also connected by vertical grooves which permit of the circulation of the liquid, and also allow any gas which may be generated to escape. The positive plate, which is made considerably heavier than the negative, is surrounded by asbestos cloth which prevents any active material, which may become loose, from falling out and causing short circuits between the plates. The asbestos cloth, it will be noted, encircles the bottom of the

Fig. 89.

plates as well as the sides so that no material can fall to the bottom. Fig. 80 shows the plate complete in perspective.

The capacity of the Chloride cells is from 5 to 6 ampere hours per pound, with a discharge rate of one-half ampere for each pound of plate—a very high rate. Notwithstanding this high capacity and high rate of discharge the efficiency of the cell is very high, the loss in current being less than 10 per cent. and of watt efficiency from 75 to 85 per cent. Experiments have shown that at the rate of one-half ampere over three-quarters of the capacity is obtained above 2 volts. For electric launch work this feature is evidently a very valuable one.

106. A very compact form of storage battery possessing, it would seem, various points of advantage in electrical boat work is that designed by Donato Tommasi, of Paris, and here shown in Figs. 90 and 90a. Each electrode is composed of a perforated tube of lead, ebonite, porcelain or celluloid, the bottom of which is closed by a plate of ebonite, in the centre of which is fixed a rod of lead, which acts as a conductor. The space between the central rod and the walls of the tube-electrode is filled with the oxide of lead. Metallic contacts connect respectively the rods of

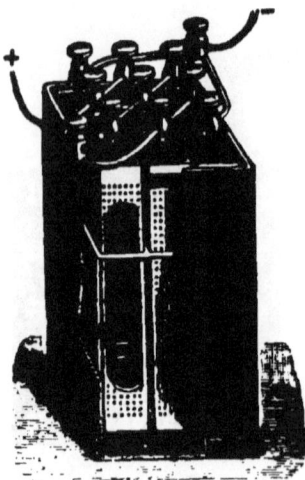

Fig. 90.—Tommasi's Multitubular Storage Battery.

the positive tubes with the rods of the negative tubes, so that the current, in order to pass from one to the other, is obliged to spread over the entire active mass and thus produce a chemical circuit without loss and with uniform action throughout the active material.

The tubular electrode from which the best results have been obtained is in the form of a rectangular cylinder, as shown in the accompanying illustration, and, in this form of the central lead rod, is provided with a number of wing-like projections. Special precautions are, of course, taken to prevent the coming in contact of electrodes of different polarity. As a result of this arrangement, the active matter, and hence the capacity of the cell, is greatly increased, and its weight is said to be from two to six times less, and its volume four to eight times less, than that of the accumulators at present in use. M. Tommasi also claims that in forming or charging his multitubular battery a current of 60 amperes per kilogramme of electrode may be employed, as against one ampere employed in present practice. On account, also, of the absence of all soldered joints in the connections between the different

FIG. 90a.—TOMMASI'S MULTITUBULAR STORAGE BATTERY.

elements, all interruptions in service are prevented. This type of cell also is free from expansion of the tube, and the active matter, being entirely enclosed, does not fall, and hence a short circuit cannot take place. The illustration, (Fig. 90a) shows a set of these cells connected up for work. The Tommasi accumulator includes 67 per cent. of active matter, the ratio of active matter to that of lead in weight being about 2.1 to 1 ; that is, for 100 grammes of lead there are 210 grammes of active matter. The following figures give the principal electrical details of the cell: E. M. F.,

2.4 volts ; capacity, per kilogramme of electrode, 16 ampere
hours; current efficiency, 95 per cent.; watt efficiency, 80
per cent.

107. A few years ago M. Emile Reynièr, of France,
brought out a cell intended to furnish a high voltage in
and of itself, particularly for such work as boat propulsion.
It is shown in Fig. 91, and consists of 16 plates mounted in
flexible pockets, so as to have a certain amount of elasticity.
These elements are placed flat, one against the other, and
compressed between two end plates of wood by means of

FIG. 91.—REYNIÈR'S ELECTRIC HIGH POTENTIAL ACCUMULATOR.

rubber spring bands. A bridge, consisting of hard wood
impregnated with a waterproofing material, carries the
whole, which may be suspended or rest upon its base, as
desired.   The spring arrangement gives to the active solid
matter an artificial elasticity, which results in large specific
power and storing capacity.   The continuous compression
of the plates, insulators and flexible pockets insures for
these thorough protection against shaking and rough
handling. Each of the pockets into which the plates are
inserted is closed on top by means of a flexible and
insulating stopper.

We give below the principal figures relative to the cell, which has 16 couples and which is known as the horse-power-hour type.

E. M. F. ....32 volts.
Available fall of potential.................28 volts.
Current discharge..................3 to 6 amperes.
Normal power, about.....................150 watts.
Capacity ....30 ampere hours.
Available useful energy............740 watt hours.
External dimensions : $\begin{cases} \text{Length} ....0.40 \text{ metre.} \\ \text{Breadth}.........0.30 \quad " \\ \text{Height}..........0.30 \quad " \end{cases}$
Contents, without containing cell.....36 cub. decim.
Total weight without cell ....50 kilogr.
Weight per kilowatt....................330   "
   "   "   kilowatt-hour.................67   "
Volume "   kilowatt.......... 240 cub. decimetres.
   "   "   kilowatt-hour 49   "   "

108. Mention may be made of the use of "Lithanode," for launch work. Its employment is due to the investigations of D. G. Fitz-Gerald, who has advocated its advantages as an active material for the anodes or positive plates. It is peroxide of lead in a dense, coherent, and highly conductive form. Its composition is almost the same as that of active material in general, but it differs somewhat in molecular construction and in its freedom from local action. The electromotive force developed by lithanode in conjunction with spongy lead is 2 volts. With a combination of lithanode and zinc, an electromotive force of 2.5 volts is obtained. The electrolyte is a solution of sulphuric acid and water, of a density about 1,220. In cells for heavy work, the elements are made up of a number of small slabs of lithanode, whose outer edges are V-shaped, and which are at once very hard and very porous. Around these, in the casting frame, molten lead or lead alloy is run, so as to fill the interstices and make a strong, complete plate out of which the pellets cannot fall. The negatives plates are built upon a corresponding plan ; and where lightness is desirable a copper gauze spongy

lead negative is used. A marked reduction of weight is
said to be obtained in all types of the lithanode cell.

109. Reference has already been made to the use of a
semi-solid electrolyte, which could not splash or spill.
Dr. Paul Schoop,[1] of Switzerland, has been an active
worker in this field, and to him we owe a successful
gelatinous electrolyte, obtained by adding one volume of
dilute sodium silicate—water glass—density 1.18 to two
volumes of dilute sulphuric acid 1.250 density. Similar
work has been done in England by Mr. Barber-Starkey,
who has interposed a mixture of plaster of Paris and

FIG. 92.

sawdust between the plates; and in America by Mr.
Pumpelly, who has thus used cellulose or wood pulp.

110. We do not know of any present instance in electrical
boat work, in which the motor is not directly connected
with the shaft of the screw, thus avoiding gearing. An
example of opposing practice is that here illustrated in
Figs. 92 and 93, which show the "Electricity" already
referred to,[1] the first electric launch, we believe, ever
propelled upon the English Thames. Fig. 93 shows two
motors connected by belts to an overhead countershaft and
arranged with a friction clutch, by means of which either
motor could be thrown in or out of gear. From the
countershaft, the belting passed down to a pulley on the

1. See Chapter VIII, page 96.
1. Chapter II, page 12.

axis of the screw. Such a method was, however, only tried to be abandoned, and we do not know of its revival.

The modern motors, of which several have already been illustrated and described in these pages,[1] may be taken as types of machines many of which are obtainable to-day, admirably suited for direct driving in boat service. One type, however, that we have not yet included, but whose shape adapts it remarkably to the narrowed, converging

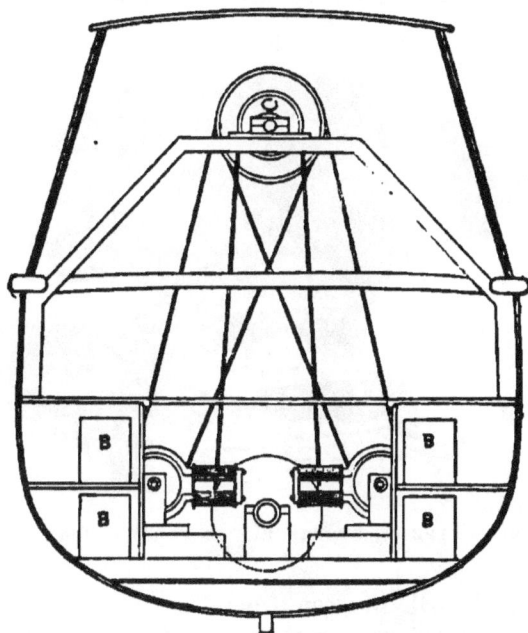

Fig. 93.

sides of a boat is the Storey, a form of which is shown in Fig. 94. It is the motor illustrated in the Chamberlain rowboat[2] where the drawing gives an excellent idea of the manner in which the sides of the boat cradle the motor, and at the same time permit it to rest very close to the keel, thus bringing the propeller shaft very low. The

1. See pages 22, 31, 34, 45.
2. See Chapter V, page 52.

motor is a perfect cylinder in shape, is ironclad, and is watertight.

111. A reversion to anterior methods has been made recently in this country, in trying again the plan of Trouvé of mounting the motor on the rudder.[1] There is now in operation on the Schuylkill River, at Fairmount Park, Philadelphia, an interesting electric launch, built by Mr. F. A. La Roche, of that city. It measures 16 feet on the water line, 3 feet 10 inches beam, and draws about 1½ feet

FIG. 94.—TYPE OF STOREY MOTOR FOR LAUNCH WORK.

of water. It will be seen from Fig. 95 that the entire mechanism is set on the stern post in exactly the same manner as a rudder is hung on an ordinary boat or launch and may be removed at pleasure and set on any other boat without alteration, provided, of course, that the rudder hinges are the same distance apart on both.

The accumulators, six in number, are of the La Roche type, each having twelve plates 6 by 6 inches and each cell weighing 25 pounds, making the entire weight of all six (encased in two wooden boxes, which may be set in any convenient part of the boat) only 175 pounds. The floor

1. Chapter I , page 4.

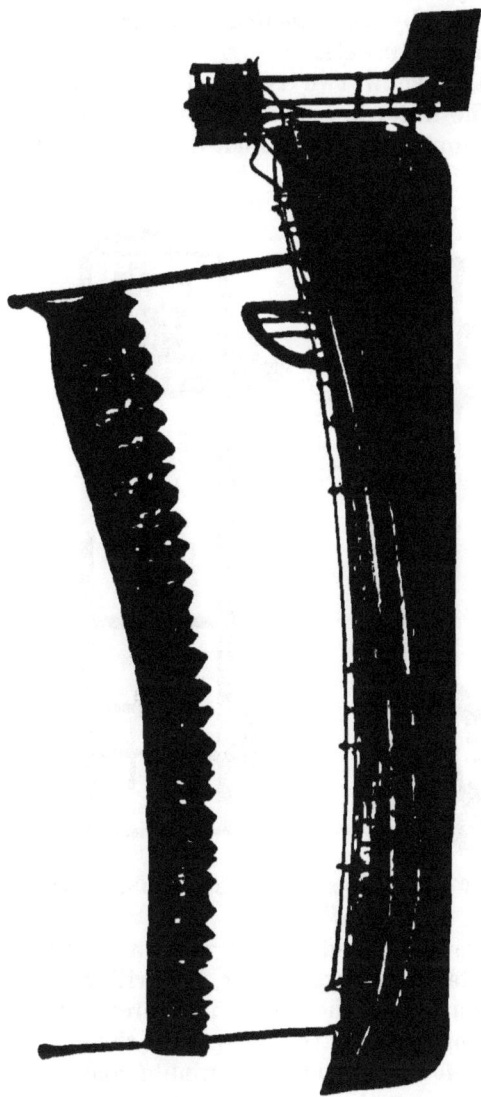

Fig. 95.—The La Roche Electric Launch in the Schuylkill River, Fairmount Park, Philadelphia.

space occupied by the boxes is exactly 3 square feet. Current is conveyed to the motor through the rheostat and a reversing switch by an ingenious arrangement of the brass railing on the gunwale.

Figs. 96 and 97 show two types of motor and their mechanism. The latter is a small ⅛ horse-power machine capable of developing in the launch described, and with

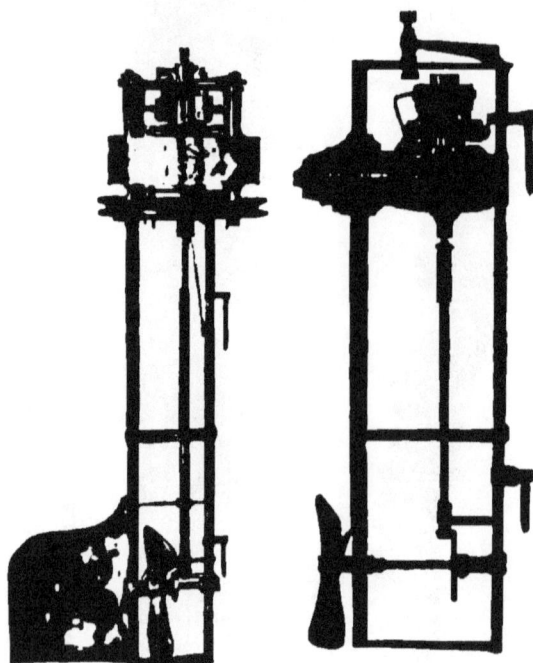

FIGS. 96 AND 97.—LA ROCHE ELECTRIC LAUNCH MOTORS.

six cells of battery, a speed of four miles an hour, while the former is rated at ¼ horse-power, but will give nearly double that amount for a short time if desired. All parts are made of aluminum, and the motor is of the multipolar type running at 400 revolutions per minute, and can drive the 16 foot launch at the rate of 8 miles per hour. In practice the motor is covered with a water-tight sheet iron cap. Mr. A. L. Riker has lately tried similar designs of motor on rudder in this country.

112. The use of a dynamotor has more than once been touched upon in these pages. It will have been evident that there are many kinds of current available for the

Fig. 98.

charging of storage batteries, such as street railway current, power circuits, central station, isolated plant, etc.,—but it is also plain that frequently the current will have to

TABLE OF DYNAMOTOR DIMENSIONS AND CAPACITY.

Size.	Dimensions Over All.			Watts Dynamo End	Shipping Weight of Transformer. Pounds.
	A. Length. Inches.	B. Breadth. Inches.	C. Height. Inches		
1/8	10 1/16	6 1/2	8 1/2	35	40
*1/8	11 1/8	6 1/2	8 5/16	45	41
1/4	15 15/16	8 3/8	10 13/16	80	124
1/2	18 1/8	10	12 15/16	175	185
1	19 1/2	13 1/4	15	500	325
2	23	14 1/2	17 3/8	1000	415
3	27 5/16	16 1/4	18 1/2	1500	590
5	29 1/2	19 1/4	21 3/8	2500	840
7 1/2	33 3/8	21 1/2	23 1/2	4250	1050
10	36	23 5/16	25 1/2	6000	1590

* Long Commutator for Large Currents of Low Voltage.

be transformed to render it suitable for charging. We illustrate here, in Fig. 98, a Crocker-Wheeler standard dynamotor, or direct current transformer, and supplement it by a table which will give an idea of dimensions. Such machines may be kept ashore or carried on the boat; their flexibility in stepping a current up or down or changing in nature, is practically unlimited; and the authors of this volume believe that their utility will be strikingly demonstrated in the field of electrical navigation, although it is to be expected that other methods will also develop as the art advances.

INDEX.

www.ingramcontent.com/pod-product-compliance
Lightning Source LLC
Chambersburg PA
CBHW021657210326
41599CB00013B/1450